# Lessons and Legacies of INTERNATIONAL POLAR YEAR 2007-2008



Committee on the Lessons and Legacies of International Polar Year 2007-2008

Polar Research Board

Division of Earth and Life Studies

NATIONAL RESEARCH COUNCIL
*OF THE NATIONAL ACADEMIES*

THE NATIONAL ACADEMIES PRESS
Washington, D.C.
**www.nap.edu**

**THE NATIONAL ACADEMIES PRESS • 500 Fifth Street, NW • Washington, DC 20001**

NOTICE: The project that is the subject of this report was approved by the Governing Board of the National Research Council, whose members are drawn from the councils of the National Academy of Sciences, the National Academy of Engineering, and the Institute of Medicine. The members of the committee responsible for the report were chosen for their special competences and with regard for appropriate balance.

This study was supported by the National Science Foundation under award number ANT-1026273. Any opinions, findings, and conclusions, or recommendations expressed in this material are those of the author(s) and do not necessarily reflect the views of the sponsoring agency or any of its sub agencies.

International Standard Book Number-13: 978-0-309-25203-4
International Standard Book Number-10: 0-309-25203-2

Cover image by Max, age 5, Seattle, WA.

Copies of this report are available from the program office:
Polar Research Board
500 Fifth Street, NW
Washington, DC 20001
(202) 334-3479

Additional copies of this report are available for sale from the National Academies Press, 500 Fifth Street, NW, Keck 360, Washington, DC 20001; (800) 624-6242 or (202) 334-3313; http://www.nap.edu/.



Printed in the United States of America

# THE NATIONAL ACADEMIES

***Advisers to the Nation on Science, Engineering, and Medicine***

The **National Academy of Sciences** is a private, nonprofit, self-perpetuating society of distinguished scholars engaged in scientific and engineering research, dedicated to the furtherance of science and technology and to their use for the general welfare. Upon the authority of the charter granted to it by the Congress in 1863, the Academy has a mandate that requires it to advise the federal government on scientific and technical matters. Dr. Ralph J. Cicerone is president of the National Academy of Sciences.

The **National Academy of Engineering** was established in 1964, under the charter of the National Academy of Sciences, as a parallel organization of outstanding engineers. It is autonomous in its administration and in the selection of its members, sharing with the National Academy of Sciences the responsibility for advising the federal government. The National Academy of Engineering also sponsors engineering programs aimed at meeting national needs, encourages education and research, and recognizes the superior achievements of engineers. Dr. Charles M. Vest is president of the National Academy of Engineering.

The **Institute of Medicine** was established in 1970 by the National Academy of Sciences to secure the services of eminent members of appropriate professions in the examination of policy matters pertaining to the health of the public. The Institute acts under the responsibility given to the National Academy of Sciences by its congressional charter to be an adviser to the federal government and, upon its own initiative, to identify issues of medical care, research, and education. Dr. Harvey V. Fineberg is president of the Institute of Medicine.

The **National Research Council** was organized by the National Academy of Sciences in 1916 to associate the broad community of science and technology with the Academy's purposes of furthering knowledge and advising the federal government. Functioning in accordance with general policies determined by the Academy, the Council has become the principal operating agency of both the National Academy of Sciences and the National Academy of Engineering in providing services to the government, the public, and the scientific and engineering communities. The Council is administered jointly by both Academies and the Institute of Medicine. Dr. Ralph J. Cicerone and Dr. Charles M. Vest are chair and vice chair, respectively, of the National Research Council.

**www.national-academies.org**

## COMMITTEE ON THE LESSONS AND LEGACIES OF INTERNATIONAL POLAR YEAR 2007-2008

**JULIE BRIGHAM-GRETTE** (*Co-Chair*), University of Massachusetts, Amherst
**ROBERT A. BINDSCHADLER** (*Co-Chair*), NASA Goddard Space Flight Center, Greenbelt, MD
**MARY R. ALBERT,** Dartmouth College, Hanover, NH
**JOHN J. CASSANO,** University of Colorado, Boulder
**LARRY D. HINZMAN,** University of Alaska, Fairbanks
**EILEEN E. HOFMANN,** Old Dominion University, Norfolk, VA
**IGOR KRUPNIK,** Smithsonian Institution, Washington, DC
**VERA KINGEEKUK METCALF,** U.S. Arctic Research Commission/Eskimo Walrus Commission, Nome, AK
**STEPHANIE PFIRMAN,** Barnard College, Columbia University, New York, NY
**CHRIS RAPLEY,** University College London, UK
**LISA SPEER,** Natural Resources Defense Council, New York, NY
**THOMAS N. TAYLOR,**[*] University of Kansas, Lawrence
**WILFORD F. WEEKS,** University of Alaska, Fairbanks (Emeritus), Portland, OR

*NRC Staff*

**MARTHA MCCONNELL,** Program Officer until September 2011
**EDWARD DUNLEA,** Senior Program Officer
**LAUREN BROWN,** Research Associate
**SHELLY FREELAND,** Senior Program Assistant

[*] Resigned during the study process.

## POLAR RESEARCH BOARD

**JAMES W. C. WHITE** *(Chair)*, University of Colorado, Boulder
**SRIDHAR ANANDAKRISHNAN,** Pennsylvania State University, University Park
**JULIE BRIGHAM-GRETTE,** University of Massachusetts, Amherst
**DAVID H. BROMWICH,** Ohio State University, Columbus
**JENNIFER A. FRANCIS,** Rutgers University, Marion, MA
**EILEEN E. HOFMANN,** Old Dominion University, Norfolk, VA
**BERNICE M. JOSEPH,** University of Alaska, Fairbanks
**AMY LAUREN LOVECRAFT,** University of Alaska, Fairbanks
**MOLLY E. McCAMMON,** Alaska Ocean Observing System, Anchorage
**ELLEN S. MOSLEY-THOMPSON,** Ohio State University, Columbus
**GEORGE B. NEWTON,** QinetiQ North America and U.S. Arctic Research Commission, Marstons Mills, MA
**CARYN REA,** ConocoPhillips, Anchorage, AK
**VLADIMIR E. ROMANOVSKY,** University of Alaska, Fairbanks
**GAIUS (GUS) R. SHAVER,** The Ecosystems Center, Woods Hole, MA
**ALLAN T. WEATHERWAX,** Siena College, Loudonville, NY

*Ex-Officio*

**JACQUELINE M. GREBMEIER,** University of Maryland, Solomons
**MAHLON (CHUCK) C. KENNICUTT II,** Texas A&M University, College Station
**TERRY WILSON,** Ohio State University, Columbus

*NRC Staff*

**CHRIS ELFRING,** Board Director
**LAURIE GELLER,** Senior Program Officer
**EDWARD DUNLEA,** Senior Program Officer
**LAUREN BROWN,** Research Associate
**RITA GASKINS,** Administrative Coordinator
**GRAIG MANSFIELD,** Financial Associate
**AMANDA PURCELL,** Senior Program Assistant

# Preface

Despite their location, tucked away at the fringes of maps of our planet, the polar regions are central to the global system. Scientists of the International Geophysical Year in 1957 could not have imagined the extent to which humanity has changed the face of our planet in the intervening 50 years. Record lows in the extent of Arctic summer sea ice, rapid changes in the Greenland ice sheet, the disintegration of gigantic ice shelves around the Antarctic, ocean acidification, and reorganization of polar ecosystems, among other changes, are reshaping the world, which is now home to over 7 billion people.[1]

What we primarily celebrate in this International Polar Year 2007-2008 (IPY) are the scientific pursuits that illuminate our understanding of the high latitudes and the role that they play in a rapidly evolving world. Reaching across the scientific spectrum, from the first high-resolution images of whole mountain ranges buried beneath Antarctica to the asymmetric auroras of our austral and boreal atmosphere, IPY 2007-2008 focused attention on the Earth as a complex integrated system. New technologies, new tools, and networked data acquisition structures were developed, setting new benchmarks for observing and understanding polar systems. Our understanding of the risks and uncertainties of global change were enhanced through groundbreaking modeling studies of the geologic past.

Starting from the efforts of a small number of enthusiasts and agencies, and building on existing multinational collaborations and science programs, IPY developed into a worldwide, community-based effort. Central to this success was an expanding Internet that permitted the rapid growth of a community; the transmission of ideas, maps, and data; the matching of collaborators; and the evolution of innovative themes. The Internet also made it possible for scientists to engage the public personally and enter thousands of classrooms as never before, often directly from remote field sites, as one part of the larger IPY education and outreach effort. IPY also celebrated the human spirit of discovery, bridging circumarctic indigenous knowledge with shared scientific endeavors while also addressing challenging societal concerns.

At its core, IPY was a large, coordinated suite of polar observations, research, and analysis. It also achieved an expanding knowledge base of diverse and enthusiastic men and women prepared to sustain and build on the legacies of previous polar science. Many dedicated people deserve thanks for their efforts in this process. IPY would not have happened but for the dedication and efforts of the thousands of participating scientists and researchers. Many more technicians and engineers assisted science teams with equipment and logistics in challenging environments.

This report was prepared to capture the major successes of this effort and to summarize what was learned. The committee heard from many people in the polar science community, and we thank everyone for their thoughts and perceptions (see Acknowledgments section). On behalf of the entire study team, we also thank the National Science Foundation's Office of Polar Programs for their support of IPY and this

[1] *http://www.census.gov/.*

report, and for providing documentation and informative details. Finally, this report would not have been possible without the dedication and hard work of the National Research Council staff: Martha McConnell, Shelly Freeland, Lauren Brown, Edward Dunlea, and Chris Elfring.

The world will continue to change, and processes of polar amplification will continue the rapid transformation of the high latitudes in the coming decades. Our hope is that the legacies of IPY will help societies understand those changes and put knowledge into action, forging new frontiers in the protection and management of our planet's resources at all latitudes. When another IPY is needed in the future, we hope the lessons from this one can serve as a guide.

Julie Brigham-Grette,
University of Massachusetts, Amherst
Robert Bindschadler,
NASA Goddard Space Center
Co-Chairs
Committee on the Lessons and Legacies
of International Polar Year 2007-2008

# Acknowledgments

This report has been reviewed in draft form by individuals chosen for their diverse perspectives and technical expertise, in accordance with procedures approved by the National Research Council's (NRC's) Report Review Committee. The purpose of this independent review is to provide candid and critical comments that will assist the institution in making its published report as sound as possible and to ensure that the report meets institutional standards for objectivity, evidence, and responsiveness to the study charge. The review comments and draft manuscript remain confidential to protect the integrity of the deliberative process. We wish to thank the following individuals for their review of this report:

**Robin Bell**, Columbia University, Palisades, NY
**Cecilia Bitz**, University of Washington, Seattle
**Richard Boone**, University of Alaska, Fairbanks
**David Carlson**, IPY International Programme Office, Boulder, CO
**Shari Gearheard**, National Snow and Ice Data Center, Clyde River, Nunavut, Canada
**Guy Guthridge**, National Science Foundation (Retired), Arlington, VA
**Amy Lovecraft**, University of Alaska, Fairbanks
**Mark Parsons**, National Snow and Ice Data Center, Boulder, CO
**Lynne Talley**, Scripps Institution of Oceanography, La Jolla, CA

Although the reviewers listed above have provided constructive comments and suggestions, they were not asked to endorse the report's conclusions or recommendations, nor did they see the final draft of the report before its release. The review of this report was overseen by Margo H. Edwards, University of Hawaii at Manoa. Appointed by the NRC, she was responsible for making certain that an independent examination of this report was carried out in accordance with institutional procedures and that all review comments were carefully considered. Responsibility for the final content of this report rests entirely with the authoring panel and the institution.

In addition, the committee would like to thank in particular for their contributions during the study process:

Waleed Abdalati, National Aeronautics and Space Administration, Washington, DC
Richard Alley, Pennsylvania State University, University Park
Jenny Baeseman, Association of Polar Early Career Scientists, Tromsø, Norway
Robin Bell, Lamont-Doherty Earth Observatory Columbia University, New York, NY
Jody Deming, University of Washington, Seattle
Hajo Eicken, University of Alaska, Fairbanks
Jackie Grebmeier, University of Maryland, College Park

Geoffrey Haines Styles, POLAR-PALOOZA/P2K, New York, NY
Fae Korsmo, National Science Foundation, Washington, DC
Don Perovich, ERDC-CRREL, Hanover, NH
Ted Scambos, NSIDC, Boulder, CO
Peter Schlosser, Lamont-Doherty Earth Observatory, Columbia University, New York, NY
Carthage Smith, International Council for Science, Paris, France
Mead Treadwell, Lt. Governor of Alaska, Juneau

The committee would also like to thank the numerous scientists spoken to throughout the study process, in particular all of the questionnaire respondents who provided their thoughts on the lessons and legacies from IPY.

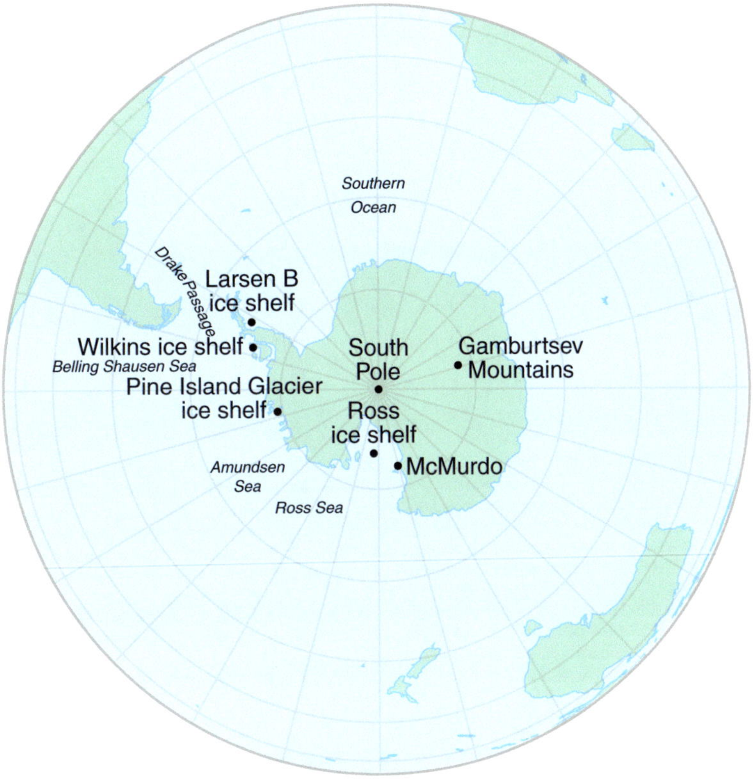


Map of Locations in the Antarctic Described in the Report

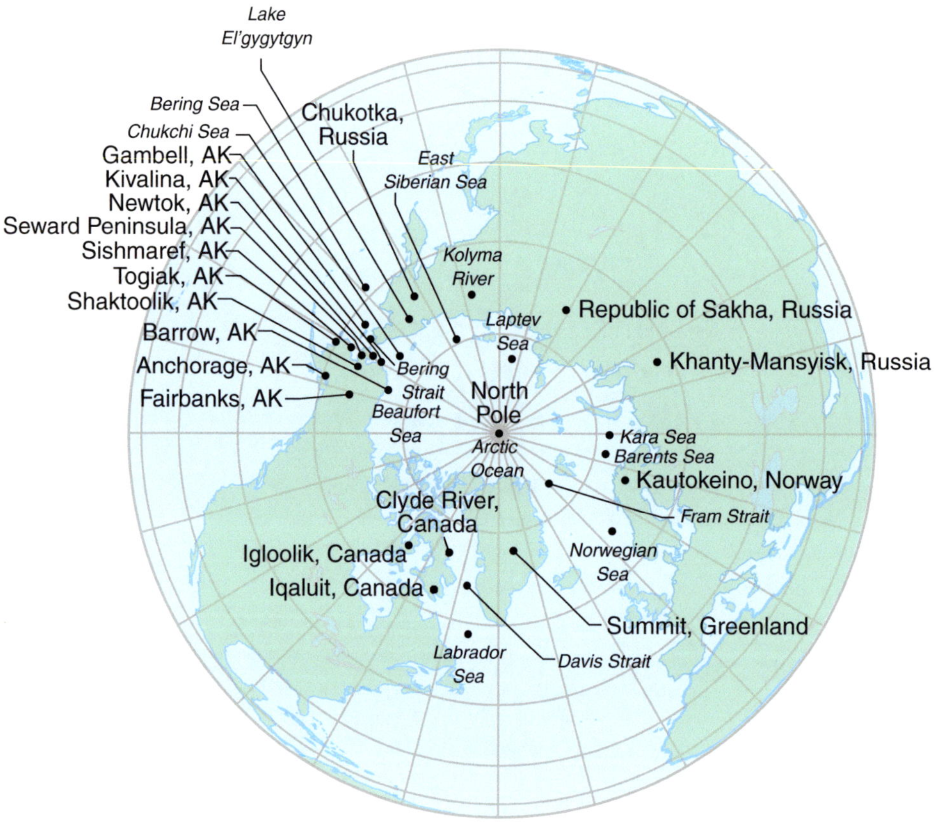


Map of Locations in the Arctic Described in the Report

# Contents

# Summary

Initiated by a core group of enthusiasts and agencies with knowledge of previous international polar years, and building on existing international programs and organizational infrastructure, International Polar Year 2007-2008 (IPY[1]) grew from the ideas and grassroots efforts of polar scientists around the world. The impetus lay in emerging evidence of the importance of the polar regions in the global system; the timing presented an opportunity both to celebrate the 50th anniversary of the International Geophysical Year (IGY) in 1957-1958 and to match or exceed the significant and enduring scientific contributions of the IGY. With its scientific focus on an integrative understanding of the polar regions in a time of rapid planetary change, IPY was the right initiative at the right time. At a time when the polar regions, and in particular the Arctic, are undergoing a transformation from an icy wilderness to a new zone for human affairs, the insights afforded by IPY could not be more timely or relevant.

Like its predecessor initiatives, IPY was designed to be an intense, coordinated field campaign of polar observations, research, and analysis. It was planned to facilitate both individual and national involvement and to allow scientists and agencies to focus on their priority issues through national peer review funding processes, ensuring cutting-edge science.

IPY attracted the involvement of more than 60 nations (Krupnik et al., 2011) and brought attention to a broad array of compelling interdisciplinary science. It was championed by ICSU[2] and WMO,[3] which provided key international nongovernmental and intergovernmental endorsement. An estimated 50,000 researchers, local observers, educators, students, and support personnel were involved in the 228 international IPY projects and numerous related national efforts.

Early indications of the IPY results demonstrate that the functioning of the Earth system cannot be understood without knowledge of the state and dynamics of the polar regions. The focused attention of IPY provided a forum to help people understand that the polar regions matter to all life on Earth. As humanity grapples with the complexities and challenges of changes in the environment and in societies around the world, the lessons and legacies from IPY offer information and inspiration for decision makers and planners now and in the future.

IPY was the largest, most comprehensive campaign ever mounted to explore Earth's polar domains. Under its auspices, scientists worked together to unlock the secrets of the Arctic and Antarctic: How does life persist in these coldest, darkest corners of the globe? How will changes in glaciers, ice sheets, snow cover, and sea ice affect the Earth system? How are traditional ways

[1] Throughout this report, International Polar Year 2007-2008 is referred to as IPY or IPY 2007-2008.

[2] ICSU, the International Council for Science, is a nongovernmental organization with a global membership of national scientific bodies (121 members, representing 141 countries) and international scientific unions (30 members).

[3] WMO, the World Meteorological Organization, is a specialized agency of the United Nations for meteorology (weather and climate), operational hydrology, and related geophysical sciences with a membership of 189 member states and territories.

of life in the North facing the challenges of a changing planet? What will be discovered when 21st century technology examines this unique frontier?

Given the size and scope of IPY, it is important to ask: Was it a success? What was learned? And what could be done better next time? This report is an attempt to answer these questions by considering the accomplishments and lessons learned through IPY. Because science funding for IPY projects in the United States was awarded from 2006 to 2009, all polar science conducted during this time is recognized under the umbrella of IPY in this report on U.S. lessons and legacies.

Evidence to date shows that IPY accomplished its goals. Activities at both poles led to scientific discoveries that provided a step change in scientific understanding and yielded insights about the importance of the polar regions. IPY facilitated a major expansion of the polar science capabilities of people, tools, and systems; it inspired the engagement of educators, students, polar residents, and the public at large; and it saw the transitioning of its scientific knowledge to policy-relevant information.

## OBJECTIVES OF IPY

The International Polar Year of 2007-2008 was built on a foundation laid by International Polar Years in 1882-1883 and 1932-1933 and the International Geophysical Year in 1957-1958. In its time, each of these campaigns marked a breakthrough in internationally coordinated exploration of Earth and space. IPY 2007-2008 took place in a different context from previous such efforts. The years leading up to it saw a mounting recognition of increasing global temperatures, rising sea level, and environmental change. Events such as the breakup of the Larsen B ice shelf in Antarctica and the diminishing sea ice and seasonal opening of the Northwest Passage in the Arctic highlighted the rapid pace of change at the poles.

To address these and other polar and planetary interactions and changes, the following objectives were articulated for IPY in the 2004 NRC report *A Vision for International Polar Year* (NRC, 2004):

> • The U.S. scientific community and agencies should use the IPY to initiate a sustained effort aimed at assessing large-scale environmental change and variability in the polar regions.
>
> • The U.S. scientific community and agencies should include studies of coupled human-natural systems critical to societal, economic, and strategic interests in the IPY.
>
> • The U.S. IPY effort should explore new scientific frontiers from the molecular to the planetary scale.
>
> • The International Polar Year should be used as an opportunity to design and implement multidisciplinary polar observing networks that will provide a long-term perspective.
>
> • The United States should invest in critical infrastructure (both physical and human) and technology to guarantee that IPY 2007-2008 leaves enduring benefits for the nation and for the residents of northern regions.
>
> • The U.S. IPY program should excite and engage the public, with the goal of increasing understanding of the importance of polar regions in the global system and, at the same time, advance general science literacy in the nation.
>
> • The U.S. scientific community and agencies should participate as leaders in International Polar Year 2007-2008.

## THE HUMAN ELEMENT IN IPY

People were the engine that powered IPY. The capability, enthusiasm, and experience of the international polar research community grew through participation in IPY and the community grew more connected as participants collaborated on IPY international projects. Young polar researchers from around the world were drawn to polar science and formed an active peer network that will help empower the next generation of polar scientists. In addition to growing in number, the polar research community grew more diverse, in particular with more women becoming involved and taking leadership roles both in planning and in conducting field programs.

IPY also drew polar residents, in particular those from indigenous communities in the Arctic, into the research community and spurred partnerships in polar observations and resource management. Arctic residents became more aware of the advantages to be gained by using the outputs of scientific investigations to assist in their daily lives, and the research community enhanced its ability to return meaningful value-added products to residents. Furthermore, engagement with the inhabitants of the Arctic has led to new capacities for learning about the social processes and health of the people who live in the polar regions.[4] IPY showed that their traditional knowledge can enhance understanding of the global processes, and that science and scientists can provide effective means of achieving international discourse and penetrating boundaries.

Beyond the polar research community, why should the vast majority of people, who live in warmer areas of the planet, care about the polar regions or about IPY? The answer lies in a host of global connections and the siting of information about this planet that is unavailable anywhere else. The polar regions are essential links in the global climate system; they are growing in economic and geopolitical importance, and they hold unique information about Earth's climate history that can help scientists understand environmental changes in the context of past changes. This understanding can in turn support informed choices for the future of the planet and its inhabitants. For these and other reasons, public interest in the poles is high, even as public trust in science has declined.

The critical "what happens at the poles affects us all" message was delivered to a wide audience during IPY through a broad spectrum of outreach and education activities. Professionally produced presentations engaged audiences with big-screen videos, vivid images, compelling music, and opportunities for direct interaction with dynamic polar researchers and Arctic residents with personal stories to tell. A key element was direct involvement of the scientists—while polar research is inherently appealing, the enthusiasm and dedication of individual researchers are the best hook for communicating polar research.

U.S. outreach activities took place at museums, science centers, and schools across the country. The polar research community reached teachers with new ways of communication, and teachers proved receptive to increased availability of polar science materials.

## SCIENTIFIC ADVANCES AND DISCOVERIES

The poles are complicated, interlinked systems that are integrally connected to the rest of the planet. Study of the polar system includes knowledge of glaciology, atmospheric sciences, geosciences, space sciences, oceanography, biology, ecology, and social sciences.

During IPY, understanding of the complexity and interconnectedness of polar systems grew. The unanticipated activity of subglacial hydrological systems and their effect on ice sheet flow was revealed. Warm ocean currents were shown to have a greater impact on ice sheet behavior than previously realized. Researchers improved understanding of how reductions in sea ice and resulting changes in albedo have major implications for the amplification of polar warming and change, with impacts on the weather and climate of lower latitudes. The warming and freshening of the water in the Arctic Basin is increasingly affecting both sea ice reduction and basin stratification. For the living creatures of the polar oceans, recent work demonstrates that climate change is having a measurable effect at all trophic levels, from microorganisms to top predators. Terrestrial research has shown that warming over the land, the decline of sea ice, and the greening of the Arctic are all linked, and this observation of contemporary processes is supported by paleoclimate work on terrestrial and marine systems. In the Antarctic there emerged evidence during IPY for a previously unknown link between the springtime ozone hole, stratospheric cooling, and the increased strength of circulation in the Southern Hemisphere.

Scientific understanding of the connection of the polar regions to the rest of the planet also increased. A community compilation of lake sediment sequences, ice cores, and tree ring records from the Arctic borderlands demonstrated that the natural cooling trend of the last 2,000 years has been reversed by contemporary warming: the last five decades are the warmest on record, showing the influence of the rest of the planet on the Arctic. In recent winters, changing weather patterns in the eastern United States and Europe have been influenced by changing conditions in the Arctic. In both Antarctica and Greenland the contributions of individual ice sheets to global ocean volume were refined to more effectively account for the measured rate of sea level rise.

IPY-related research confirmed that the poles are changing faster than the rest of the planet. This was discovered by Arrhenius and verified 37 years ago by the first climate model of Manabe and Wetherald (1975),

[4] Unless otherwise indicated, this report focuses particularly on Indigenous Alaskan and other Arctic inhabitants. There are no permanent Antarctic residents.

and it emphasizes the importance of monitoring how the poles continue to change. IPY helped to not only illuminate the pace of change but also benchmark the status of the poles. Researchers observed, for example, that the Greenland ice sheet, parts of the Antarctic ice sheet, and Arctic sea ice show clear signs of change unprecedented in the Holocene era, although the situation is complex. Multiple studies of the ice flux, ice surface elevation, and ice mass changes all show clear and coherent signals of changes in Greenland and West Antarctica. Multiple independent satellite data sets show that the ice sheets are losing significant mass at increasing rates in some locations, while elsewhere—as predicted for a warmer and hence moister atmosphere—snowfall has increased. Arctic sea ice loss in recent years has been dramatic, with record minima in areal extent for 2007 and in volume for 2011, far exceeding the pace predicted by many recent models.

There has also been a new realization that the total belowground carbon pool in Arctic permafrost is more than double the atmospheric carbon pool and three times larger than the total global forest biomass, providing an additional potential positive feedback in the global system with both the release of carbon dioxide and methane from once frozen reservoirs. Paleoclimate data now show that earlier interglacial periods over the last 3.5 million years, which were warmer than today, included the repeated collapse of the West Antarctic ice sheet and likely included large reductions in the size of the Greenland ice sheet. This finding emphasizes that continued warming of the planet will cause continued ice loss and rising oceans, making coastal regions increasingly vulnerable to flooding.

The poles have long been at the frontier of exploration, and they are still places of discovery of the fundamentally new. Observation systems have a supporting role in discovery science—scientists often learn new things simply by looking. An entire mountain range under the Antarctic ice sheet, the Gambertsev Mountains, discovered during the IGY, was mapped during IPY with new, sophisticated radar methods revealing its surprisingly rugged alpine character. The Landsat Image Mosaic of Antarctica (LIMA) IPY project produced a high-resolution mosaic image with a detailed true-color view of Antarctica: penguin rookeries were mapped almost immediately and revealed new and abandoned sites, indicating substantial changes. IPY also continued to use Antarctica's unique platform from which to peer out into the solar system to observe space weather, as well as into the cosmos beyond to probe the composition and workings of the universe. Asymmetrical auroras were observed simultaneously for the first time at both poles, altering previous notions of processes that influence solar wind and Earth's magnetic field.

## SCIENTIFIC TOOLS AND INFRASTRUCTURE

The polar regions have always presented great logistical challenges because the terrain is vast, access can be complicated and expensive, working conditions are difficult, and the areas of interest frequently cross national boundaries. Efforts to adequately observe interaction among the large-scale systems frequently require international cooperation and significant planning.

An important outcome of IPY was the development of new collaborations that enhanced scientists' observational capability in many areas of the poles, including remote areas such as East Antarctica. Observation networks such as the Sustaining Arctic Observing Networks (SAON), the integrated Arctic Ocean Observing System (iAOOS), and the Southern Ocean Observing System (SOOS) developed and/or expanded during IPY.

In addition to these collaborations, IPY saw not only the use of existing tools in new ways and in new places, but also first-time deployments of novel tools for observing the polar climate, ecosystems, and beyond. Increasing use of a system science approach necessitated new observing systems to better understand variability and change, and new tools—including sea gliders, unmanned aerial systems, and animal-borne ocean sensors—allowed for more comprehensive observations of the poles than ever before. The cost and complexity of these systems often made multiagency and/or international cooperation necessary, and the use of remotely controlled autonomous observing systems became increasingly common.

Satellite systems were a particularly effective example of collaboration among countries and agencies. IPY cannot claim credit for the generation of any new satellite missions, but it did succeed in an unprecedented set of coordinated observations from spaceborne sensors operated by multiple national space agencies. The IPY Space Task Group coordination of the Polar Snapshot was so successful that the group continues to cooperate with national space agencies for the acquisition of observations that will support sustained space-based monitoring of the polar regions and offset decreasing observational capability as many satellite systems age and fail.

Observations are of little value if they are not available to researchers. But the challenges to availability multiply as data volumes increase and the needs of interdisciplinary research extend to data of unfamiliar form and content. A number of data centers in the United States stepped up to this challenge, made data management expertise available to IPY projects, and followed through with mechanisms to receive, organize, store, and make available metadata of all types to assist researchers in locating data relevant to a wide range of scientific pursuits.

The U.S. federal agencies that funded most of the U.S. participation in IPY have specific data archive requirements stated in their initial grant awards, resulting in large national data archives. IPY could have been an opportunity to encourage funding agencies in other countries to adopt similar policies, but this was not an avenue pursued by the IPY Joint Committee.

## KNOWLEDGE TO ACTION

IPY activities sought to convert knowledge gained through scientific inquiry into societally relevant information. Extensive IPY research, particularly in human health, community vulnerability, food security, and local observations of change, was aimed at practical applications to be shared with polar communities, local agencies, and grassroots organizations in Alaska and across the Arctic.

As an example of such knowledge application, the record sea ice minimum in 2007, the first year of IPY, stimulated concerted efforts to understand its cause, project plausible future trajectories, and consider systemwide implications for the coming decades. Sea ice conditions have a direct impact on quality of life in Arctic communities, and the systems perspective advanced understanding of impacts on fishing, hunting, shipping, and coastal erosion. The information is being used in Alaska to develop new management strategies and balance native versus commercial resource needs as marine ecosystems migrate north in search of cooler waters. Arctic residents also depend on knowledge of sea ice conditions for the success of traditional hunting practices. Through community-based interactions, participants in the IPY Sea Ice for Walrus Outlook activity were able to develop and deliver meaningful information to local communities that merged measurements of atmospheric and oceanic conditions with indigenous and local observations.

Research during IPY led to the identification of new marine and terrestrial species, habitats, and ranges and greatly expanded the understanding and awareness of polar biodiversity (and invasive species). For example, the Southern Ocean Global Ocean Ecosystems Dynamics (SO GLOBEC) Program, an international multidisciplinary effort designed to examine the growth, reproduction, recruitment, and overwintering survival of Antarctic krill, has yielded key insights into the working of the Southern Ocean food web. Better understanding of Antarctic and Arctic ecosystem dynamics has in turn spurred new initiatives aimed at managing human activities in the oceans, with an eye toward protecting biodiversity and maintaining ecosystem functions as these ecosystems undergo profound transformations due to climate change.

Looking to the future, IPY-related predictive modeling will continue to play a crucial role in helping commercial enterprises, individuals, and governments assess the regional and global risks associated with melting ice, sea level rise, permafrost degradation, and other effects of high-latitude changes in a warming world. Such assessments can help inform a wide variety of decisions about the management, siting, and sustainable insurance of coastal property and infrastructure, as well as community planning and zoning, construction of ice roads, emergency preparedness, disaster response, and long-term planning for moving military, industrial, and public infrastructure (and in some cases whole villages) to higher ground.

## LESSONS AND LEGACIES

IPY embraced existing, enhanced, and new programs. To examine the breadth of work in IPY, this report is based on the committee's evaluation of polar research reports, published articles, books, formal and informal research networks, workshops, and public outreach events that resulted during the 2 years of intensive activity at both poles. After reviewing many examples of IPY research, hosting a workshop to talk directly with IPY researchers, and listening to the polar science community (e.g., in conversations with colleagues), this committee concludes that IPY was an outstanding success. It fulfilled all its primary objectives and more.

Coming at a time of rapid polar and global change, IPY investments were both pertinent and timely, enabling the science community to observe and record a reference state of the polar system. The international polar science community, with the United States as a key player, was sufficiently mature and ready to undertake and execute this large endeavor.

Comments from participants indicate that IPY was seen as a rare and special opportunity, and intensive bursts of exciting and sometimes high-risk activity contributed to its success. Deadlines and the need for international collaboration helped to focus the efforts and decision making of the science community. The intensive nature of the effort built on and integrated existing national and international programs, making the "sum greater than the parts."

IPY was broadly inclusive and collaborative, and the active participation of polar residents, educators, and young researchers helped further expand its reach. Ultimately, though, its success was due to the perseverance and hard work of a core group of researchers with a passion for polar science and the desire to communicate the centrality of the high latitudes in affecting the behavior of the Earth system.

IPY leaves a priceless legacy. The polar research community grew in numbers, skills, and knowledge during the 2 years. Researchers recognized that the required observations of the polar regions are beyond the capability of any single nation and were thus motivated to forge new relationships among many nations and to engage with Arctic residents as partners in research. New international partnerships also supported new tools and observational networks that increased the ability to detect and document the polar environment. IPY changed perceptions with new scientific insights, in particular the enhanced recognition of the connectivity of the poles to the Earth system, and did so by exploiting new technical and logistical tools and capabilities.

An objective assessment must also take stock of problems and remaining challenges. To that end the committee notes that despite valiant attempts by the IPY Data Committee and several coordinating workshops, the development and accessibility of IPY data products were hampered by a shortage of time and resources. More effective interagency coordination within and across nations, particularly in funding approval and logistics, would have been beneficial, as not all scientific research priorities received adequate support and delays in national funding processes affected abilities to coordinate field research and infrastructure sharing. The impact of IPY was very uneven across polar communities; some communities were actively engaged and informed, whereas many more had sporadic or ad hoc access to IPY information and resources. Furthermore, the sustained impact and momentum of the IPY legacy will require ongoing support from funding agencies for both the observing networks and the scientists.

In future years, scientists may look back at this IPY as they wonder whether there is merit in planning another International Polar Year. This IPY benefited from a number of important ingredients that proved fundamental to its success:

- The involvement of a well-connected and well-organized group of core proponents promoting IPY was essential. The early planning group and then the official planning committee both provided a clear vision and compelling science to justify the necessary investments.
- A small amount of early seed funding was necessary to get things going and to add legitimacy in the world scientific community; U.S. funds were provided by the National Academy of Sciences and international funds by ICSU and WMO.
- Acceptance was needed from stakeholders—funding agencies and scientists—without which IPY could not have occurred. Active involvement grew from a small core group to a large and diverse range of people over the IPY planning period. For example, dozens of program managers from virtually every part of NSF, not just the Office of Polar Programs, evaluated and funded proposals for IPY-related research.

• IPY ultimately needed significant international monetary commitments from funders. Future IPY activities could benefit from the assertive engagement of funding agencies, as early as possible, perhaps through their own international planning group. During this IPY, extensive international commitment, participation, and support provided the stimulus for new and creative collaborations in research that would otherwise have taken longer to reach fruition.

• Coordination—both within the United States and across nations—was essential to establish structures and services to support all participants and ensure that they shared a common vision of IPY and understood that they were part of a bigger whole. The International Programme Office (IPO[5]) was key in this regard.

[5] The IPO was located in Cambridge, UK, and was supported by funding from the UK and eight other nations to keep the IPY network running throughout 2005-2010.

# 1

# Introduction

International Polar Year 2007-2008 (IPY[1]) was an intense, coordinated field campaign of polar observations, research, and analysis that ran from March 1, 2007, to March 9, 2009. Following the efforts of a core group of enthusiasts and agencies, and building on existing international programs and networks, IPY grew from the grassroots efforts of polar scientists around the world. With the active involvement of more than 50,000 participants from numerous science disciplines, institutions, and more than 60 nations (Krupnik et al., 2011), IPY represented the most comprehensive and sophisticated effort ever undertaken to understand the secrets of Earth's polar domains.

IPY 2007-2008 was built on a foundation laid by the International Polar Years of 1882-1883 and 1932-1933, and the International Geophysical Year of 1957-1958 (see Box 1.1). In its day, each of these represented major internationally coordinated efforts to advance exploration of Earth and increase human understanding of the Earth system.

The planning process for the 2007-2008 IPY began with conversations and first actions around 2000-2002, initially among small groups of scientists and encouraged by organizations responsible for the coordination of polar science, such as the U.S. Polar Research Board (PRB), the Scientific Committee on Antarctic Research, the European Polar Board, and the International Arctic Science Committee. A modest investment by the International Council for Science (ICSU) in early 2003, followed by the endorsement by the World Meteorological Organization (WMO), drew international interest to the project and legitimized the process by which it was defined and organized. ICSU established a Planning Group of 14 members, who worked from July 2003 through October 2004 to produce an overarching IPY plan, a set of objectives, and a framework for action (Rapley and Bell, 2004).

**BOX 1.1**
**History of International Polar Years**

International Polar Year 2007-2008 was an ambitious program following in the footsteps of three similar programs over the last 125 years (also see time line in Figure 1.1). The first International Polar Year in 1882-1883 comprised 12 countries and 15 expeditions (13 in the Arctic and 2 in the Antarctic). The U.S. contribution included establishment of the longest-serving U.S. scientific station in the Arctic, at Point Barrow, Alaska.

The second International Polar Year in 1932-1933 had participation from 40 nations and led to advances in meteorology, atmospheric sciences, geomagnetism, and the "mapping" of ionospheric phenomena that advanced radioscience and technology. The United States established the first year-round research station inland from the Antarctic coast.

The International Geophysical Year (IGY) in 1957-1958, with the participation of 67 nations, saw many "firsts," such as the launch of the world's first satellites. IGY had a strong polar component, especially in the Antarctic where the United States established research stations at the South Pole and McMurdo. The experience in international collaboration, even during the intense political climate of the Cold War, led to ratification of the Antarctic Treaty in 1961.

[1] Throughout this report, the terms "IPY 2007-2008" and "IPY" are used interchangeably.

Additional investments by ICSU and the WMO in 2005 established support for an IPY Joint Committee of 20 members, who steered the scientific preparation, implementation, and completion of IPY from 2005 to 2010. The daily tasks of managing the international IPY activities were coordinated by the International Programme Office (IPO) in Cambridge, UK, which was funded by the United Kingdom and eight other nations, and provided the means to establish and maintain IPY networks throughout 2005-2010. Dr. David Carlson served as Director of the IPO. Other nations established national IPY coordinating offices, many of which (e.g., the Canadian IPY office) made invaluable contributions to the international coordination of IPY.

In 2003, the PRB formed the U.S. National Committee for the International Polar Year 2007-2008. This committee conducted a study to outline the U.S. vision for IPY: What questions should it address? How should it be planned? The group, chaired by Dr. Mary Albert, sought input from across the polar science community and in 2004 published an outline of the U.S. rationale and focus for IPY, *A Vision for International Polar Year* (the Vision Report; NRC, 2004). The Vision Report was instrumental in defining the potential for IPY and sparking participation by the U.S. science community and a number of agencies. The recommendations from this report are listed in Box 1.2.

The committee also worked with senior leaders in U.S. science agencies—the National Science Foundation (NSF), National Oceanic and Atmospheric Administration (NOAA), National Aeronautics and Space Administration (NASA), and U.S. Geological Survey, among others—to encourage agency participation in the U.S. components of the IPY program. In July 2004, PRB convened a workshop to promote discussions among the federal agencies, provide a forum for their representatives to identify possible scientific activities of interest, and serve as a springboard for collaborative IPY activities. Upon completion of the workshop report (NRC, 2005), U.S. National Committee responsibilities transitioned to the PRB, at the time chaired by Dr. Robin Bell. Some members of the U.S. National Committee also participated in the IPY Planning Group (2003-2004) and Joint Committee (2005-2010).

**BOX 1.2**
**Recommendations from 2004 NRC Report on *A Vision for the International Polar Year 2007-2008***

- The U.S. scientific community and agencies should use the IPY to initiate a sustained effort aimed at assessing large-scale environmental change and variability in the polar regions.
- The U.S. scientific community and agencies should include studies of coupled human-natural systems critical to societal, economic, and strategic interests in the IPY.
- The U.S. IPY effort should explore new scientific frontiers from the molecular to the planetary scale.
- The International Polar Year should be used as an opportunity to design and implement multidisciplinary polar observing networks that will provide a long-term perspective.
- The United States should invest in critical infrastructure (both physical and human) and technology to guarantee that IPY 2007-2008 leaves enduring benefits for the nation and for the residents of northern regions.
- The U.S. IPY program should excite and engage the public, with the goal of increasing understanding of the importance of polar regions in the global system and, at the same time, advance general science literacy in the nation.
- The U.S. scientific community and agencies should participate as leaders in International Polar Year 2007-2008.

SOURCE: NRC, 2004.

The White House designated NSF the lead federal agency for organizing U.S. IPY activities.[2] In this role the NSF Office of Polar Programs (OPP) interacted with the leadership of other U.S. agencies to promote IPY and plan collaborative activities. NSF funded or cofunded the planning and execution of a wide array of science and education activities in support of IPY. For example, the NSF-funded workshop on "Bridging the Poles" in 2004 brought together scientists, educators, and media specialists to define IPY goals for integrating research, education, and outreach at the national and international levels and to build a coherent and exciting public presence during IPY (Pfirman et al., 2004). In 2005, NOAA and NSF jointly funded a workshop, "Poles Together: Coordinating International Polar Year

[2] NSF also supports the U.S. government IPY website: *http://ipy.gov/*, which contains extensive information on federally supported IPY research and activities.

(IPY) Outreach and Education" held at the University of Colorado, to develop a plan for achieving IPY goals for education, outreach, and communication (CIRES and NOAA, 2005).

## CONTEXT IN WHICH IPY TOOK PLACE

Polar research scientists who saw the potential in organizing another "international year" at a time of great planetary change were likely the biggest driver behind IPY. They realized that as global temperatures have risen (Figure 1.1), the poles are changing first and fastest and there was the need for a major campaign to increase the capacity to assess and understand the changes (Albert, 2004).

The authors of the *Arctic Climate Impacts Assessment: Scientific Report* (ACIA, 2005) also laid out a multifaceted perspective, projecting changes in every sector from sea ice and glaciers to reindeer foraging. The projected summer opening of the Arctic Ocean raised awareness of increased access to resources and potential economic and boundary disputes. The science community knew, too, that technology had changed dramatically in the 50 years since the IGY and offered rich opportunities such as satellite and airborne remote sensing and genetic sequencing (Carlson, 2011).

During the IPY planning stages, the greater scientific community began to view the poles as both a harbinger of change and a key component in the global system. In addition, as IPY was ramping up from 2004 to 2006, many members of the public, hearing of changes in the poles and other regions, began to take the issue of climate change more seriously. As an example, in 2007 the Intergovernmental Panel on Climate Change and Al Gore received the Nobel Peace Prize for raising concern about climate change. This increase in awareness of climate change was not sustained, however. Around 2006, public opinion began to shift and concern about global warming waned. In the next several years, more than 40 percent of the American public grew to feel that the seriousness of global warming was exaggerated (Figure 1.2). Also during this period before IPY, the U.S. government signed but did not ratify the Kyoto Protocol. IPY thus came at a time in the United States of growing political tension regarding climate change.

## WHAT DID IPY ACCOMPLISH?

IPY took place at a crucial time for polar and global climate change and helped to deliver the message that what happens at the poles affects all life on Earth. Many activities involved international contributors, and many emphasized societal implications, a new focus that included educational facets and ways to inform policy decisions. IPY garnered substantial support from science educators and enhanced interest from the public. From outreach activities that engaged the general

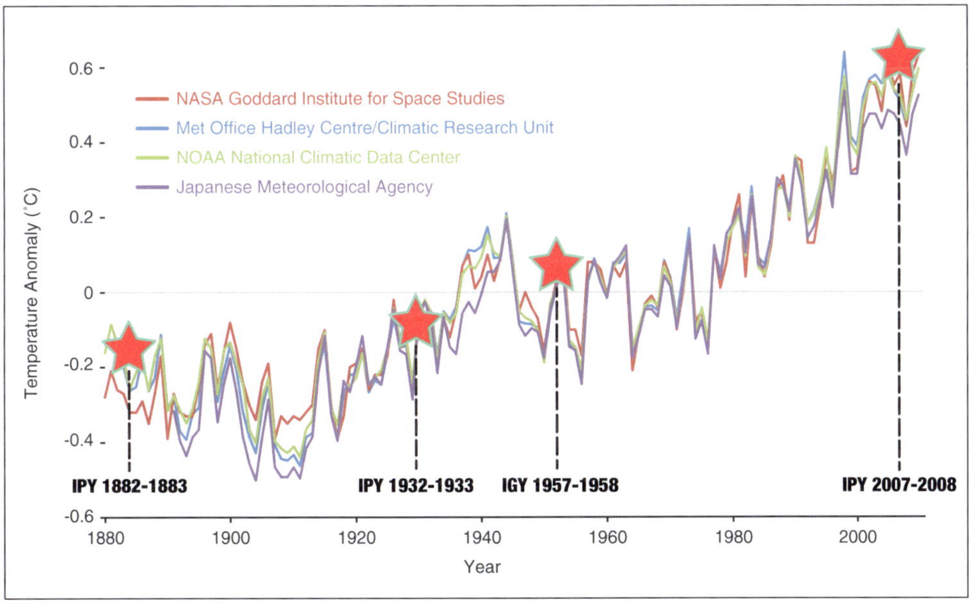


FIGURE 1.1 Global surface temperature from historical records, including dates of previous International Polar Years (IPYs) and the International Geophysical Year (IGY) SOURCE: Adapted from NASA Earth Observatory/Robert Simmon.

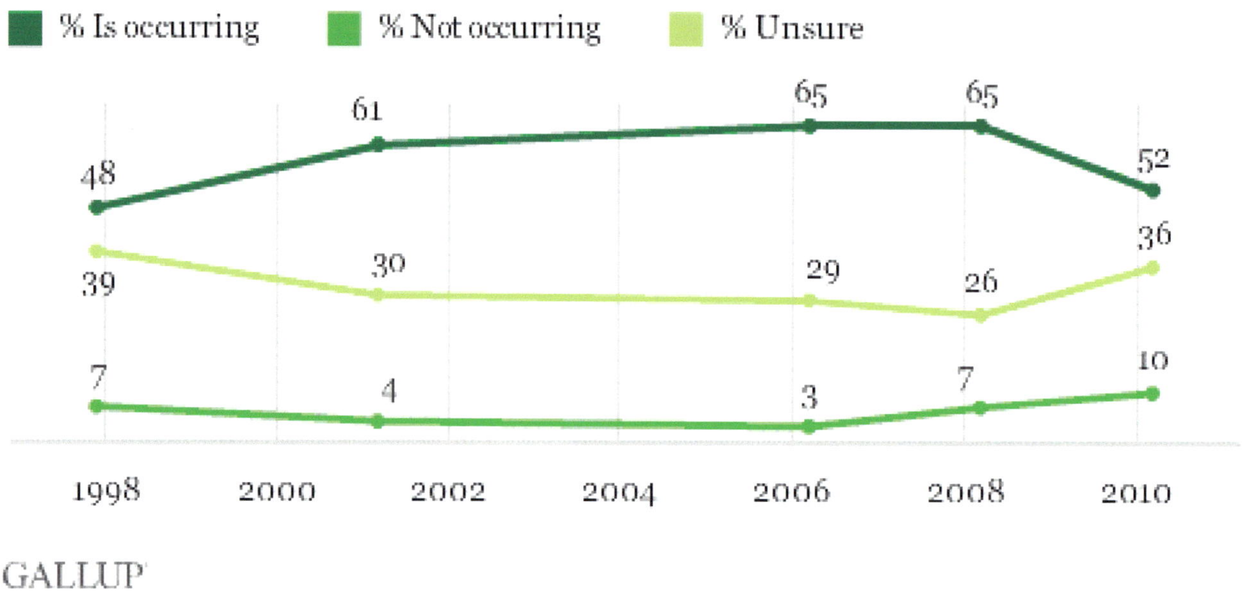


FIGURE 1.2 Shifting public opinion on global warming. Opinion polling in America suggests how people became more sure that climate scientists believed in global warming over the period 1998-2006, but more recently that level has declined for a variety of reasons, including the state of the economy, shifting media attention, and other factors (Leiserowitz et al., in press). The question asked which one of the following statements do you think is most accurate—most scientists believe that global warming is occurring, most scientists believe that global warming is not occurring, or most scientists are unsure about whether global warming is occurring or not? SOURCE: Gallup.

public to collaborative studies with indigenous people and projects that brought together researchers from multiple disciplines and many nations, the legacies of IPY are larger than its scientific results.

During IPY a major transformation occurred in the perception of the poles—from the 20th century image of them as icy, white, pristine, and uninhabited landscapes to a recognition that the poles are key, interconnected components of the Earth system and bellwethers of change, and they are thus of direct relevance to the entire globe. Among other views shared with the committee, scientific researcher Ted Scambos (University of Colorado's National Snow and Ice Data Center) offered his perceptions of the impacts of IPY:

> International Polar Year 2007-2008 represents the culmination of a transition in the way humanity views the polar regions of Earth. What was once remote and inaccessible, romantic, and challenging, is now seen as an integral part of a changing planet. This change in perception extended beyond scientists to policymakers and the public, and in large part is attributable to IPY. The previous IPYs were about exploration, and pushing fledging aspects of science (geography, geology, geophysics, space physics) further than they had gone before. While IPY 2007-2008 retained that spirit, at a fundamental level its aim was toward integrating the poles and polar systems to the rest of a changing world.

IPY planning and implementation comprised existing and enhanced scientific projects and programs as well as new initiatives, all of which fed into each other. This analysis of U.S. IPY lessons and legacies therefore encompasses the period from 2006 to 2009 because all related activities, whether ongoing or newly launched, were directly or indirectly affected by the fact that IPY was under way. Research and education activities that began before IPY and continued during it provided a foundation of established research. They also benefited from IPY because they were integrated into research meetings and education and outreach activities, often with media attention.

Other U.S. gains from IPY included significant advances in the ways U.S. science is carried out in the polar regions. Scientists in the United States already had many international connections, and the U.S. IPY program clearly benefited from and built on these collaborations.

In addition, Congress appropriated $60 million specifically for IPY, and many existing programs and resources were placed under the umbrella of IPY. This had the national benefit of facilitating their enhancement and leveraging funding through increased national and international collaboration and coordination. Similarly, sources of private funding were mobilized during IPY, in some cases independently and fortuitously. Such was the case with the establishment in 2009, by the Tinker Foundation and with support from the Sloan Foundation for the Census of Marine Life, of the Martha Muse Prize for early to mid-career investigators in Antarctic science (NRC, 2008).

In terms of scientific knowledge and understanding, IPY revealed how dynamic the Arctic and Antarctic are. Although once considered slow to change, it became clear that the poles are transforming quickly, often faster than predicted by the best models. Notwithstanding some marked contrasts between the Arctic and the Antarctic, the interconnectedness of the polar lands, oceans, ice, and human systems is evident from the data. The connections of polar systems to the global physical and human environments are increasingly obvious—for example, melting glacier ice raises sea level worldwide. Polar ecosystems may exert strong controls on global concentrations of greenhouse gases, for example, from the release of carbon dioxide and methane from thawing permafrost and subseabed methane hydrates. More discoveries are sure to follow as researchers continue to analyze data recorded during the IPY time frame.

## THE SCOPE OF THIS REPORT

In 2010, at the request of NSF OPP, under the auspices of the National Research Council, the Committee on Lessons and Legacies of the International Polar Year 2007-2008 was asked to highlight the outcomes of IPY from a U.S. perspective, integrate the lessons from different activities, and record U.S. IPY efforts so they are available to a broad audience including researchers, decision makers, and stakeholders. (The committee's Statement of Task is in Appendix A.) This report by the committee describes U.S. contributions to IPY in the context of the international breadth of IPY activities, with the goal of illustrating what has been achieved through this international-year approach and the importance of IPY to polar science and beyond.[3]

In gathering information for this report, the committee held a workshop in June 2011 at which more than 70 leading researchers, predominantly from the United States, were invited to share their findings from and perspectives on IPY. The committee also devised an online questionnaire, announced on various distribution lists used by polar researchers and IPY participants, to reach out to the polar scientists and educators. Several dozen responses to the questionnaire were received, which the committee considered as input for this report rather than as a systematic community survey. The committee sought quantitative data where possible for evaluating IPY projects and programs, but in many instances those data do not exist, so the committee relied on its own knowledge and judgment as well as its extensive information-gathering efforts. Finally, while the committee strove to maintain balance between the Arctic and Antarctic throughout the report, the United States has territory and vested interests in the Arctic, and there is therefore a natural tendency to emphasize the Arctic in certain subject areas.

Based on the feedback received, the committee judges that IPY achieved its goals of new scientific knowledge and insights. The committee members observed that, as in many other nations, U.S. IPY activities did not always strictly hew to the ICSU-WMO goals, but they nonetheless contributed significantly to priority national goals for polar science. Overall, the committee concluded that IPY expanded polar science capabilities in terms of the size and capability of the polar research community (including new research partners from nations not previously active in polar research), research tools, and systems, and it inspired educators, students, polar residents, and the public at large.

This report is structured to reflect the important facets of IPY. Because science is never accomplished without people, Chapter 2 concentrates on the human

[3] The committee was asked to address these high-level questions rather than to create a catalog of all IPY projects. The committee had to make many choices and emphasizes that the examples in this report are illustrative only and that nothing is implied by omission.

element in IPY. At its core, IPY was about polar research, so Chapter 3 describes significant scientific advances and discoveries during this period. Chapter 4 addresses the impact of IPY on tools used for polar research, and Chapter 5 focuses on the all-important need to translate scientific knowledge into actionable information. These chapters largely focus on the successes of IPY. The committee hopes that the final chapter, with reflections on the entire IPY process and experience, will prove useful to those interested in extending the series of international polar years in the future.

# 2

# The Human Element in International Polar Year 2007-2008

The wide-ranging and significant achievements of International Polar Year 2007-2008 (IPY) required the enterprise and commitment of myriad scientists, students, educators, local residents, logistics staff, program managers, and supporters—an estimated total of 50,000 people worldwide. They planned and executed the IPY programs and their direct interactions with the stakeholders and the general public brought IPY to life.

From the beginning, a major objective of IPY was to invest in "people"—that is, to expand human capacity in the quest for new scientific knowledge. This goal included increasing the numbers of current and future polar researchers (Figure 2.1) and integrating stakeholders in polar research, particularly polar residents.

A further major objective, strongly stated in the U.S. IPY Vision Report (NRC, 2004), was the creation of new connections between science and the public. The aim was effective communication of the physical and social polar sciences to increase understanding of the function of the poles in global systems. Efforts to achieve this goal engaged scientists, educators, and the media through a variety of innovative education and outreach programs.

## EXPANDING THE POLAR RESEARCH COMMUNITY

U.S. and international polar scientists represent only a small fraction of the broader scientific community, even in the geophysical sciences.[1] But the rapid and dramatic changes in the polar regions sparked both concerns about the future of the planet and the inquisitiveness of scientists from many disciplines, attracting them to investigate the many challenging scientific questions associated with these changes. The result was a measurable increase in the number of scientists conducting polar research.

One indicator of growth during and immediately after IPY is the increase in U.S. and international membership of the International Glaciological Society (IGS; Figure 2.2). IGS represents scientists who research ice in any form (including at mid- or low latitudes as well as interplanetary ice), but the overwhelming majority are involved in polar research.

A similar trend is evident in the membership of the Cryospheric Sciences Focus Group, one of the newest in the American Geophysical Union (AGU) (Figure 2.2). In 2003 and 2004, the AGU convened multiple sessions related to IPY at its annual meeting to engage the community and communicate IPY planning in the United States.

IPY contributed to a growing trend toward international collaboration for polar science, thus marking

[1] There were 1,057 members of the American Geophysical Union who in 2010 identified the Cryospheric section of AGU as their primary affiliation. Since 2001, membership in this section has increased slightly faster than the AGU membership as a whole. Cryospheric section members were 1.2% of AGU membership and are now 1.7% of total AGU membership as of 2011 (Anne Nolin, Oregon State University, personal communication).

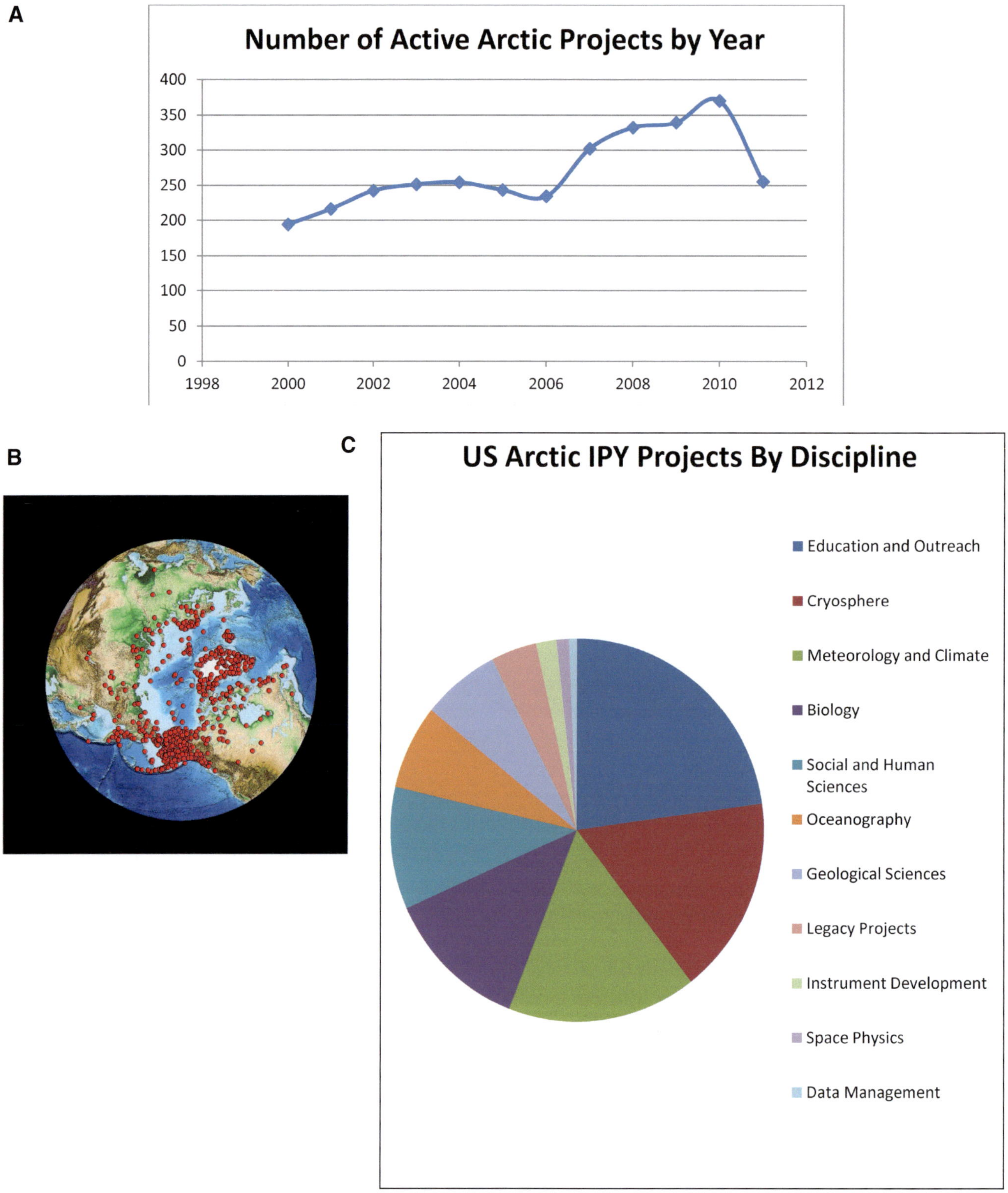


FIGURE 2.1 Data from ARMAP (Arctic Research Mapping Application) showing (A) the total number of NSF-funded field projects in the Arctic between 2000 and 2011, (B) the geographical distribution of those projects, and (C) the breakdown by discipline of the projects identified as IPY projects (of the 1,407 projects between 2000 and 2011, 188 were recognized by the National Science Foundation [NSF] as IPY projects). Most disciplines experienced a pulse of activity during IPY, especially in education and outreach; data management; and legacy projects, but there is evidence of a recent decline in the number of active projects in many regions and disciplines. Note that the ARMAP database includes only NSF-funded Arctic projects with a field-based component, so not all modeling or remote sensing projects are included. These graphics are intended to be representative of IPY efforts; they do not show the complete data for all of IPY. SOURCES: Craig Tweedie, University of Texas at El Paso; and *http://www.armap.org*.

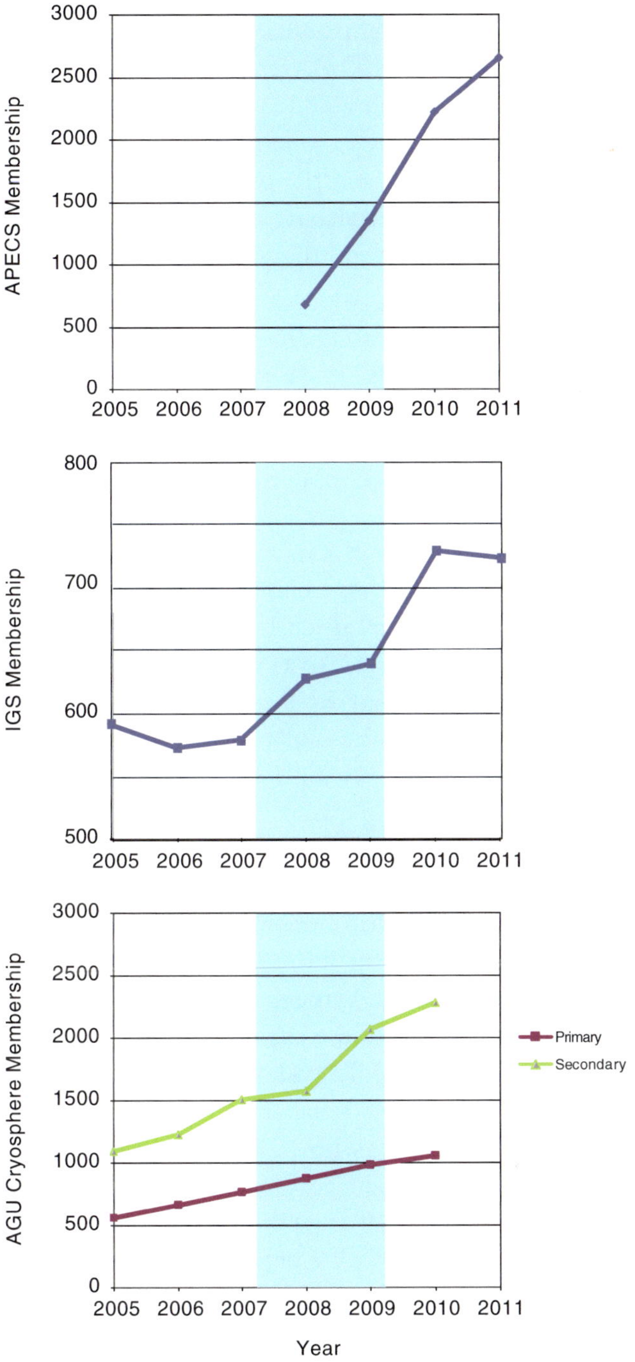


FIGURE 2.2 The membership of polar professional associations grew during IPY. Shaded region represents the official time period of IPY (March 2007 to March 2009). Top: Association of Polar Early Career Scientists (APECS) membership. Middle: Membership of the International Glaciological Society (IGS); the 2011 value includes members through August 2011. Bottom: Membership of the Cryosphere Focus Group of the American Geophysical Union. SOURCES: Data from Jenny Baeseman, APECS (top); Magnús Magnússon, IGS Office (middle); Anne Nolin, Oregon State University (bottom).

a radical departure from the previous polar/geophysical years, when most of the efforts were national in scope, logistics, and funding. The United States has been a global leader in polar research for years (Aksnes and Hessen, 2009)[2] and is by far the largest contributor to research in both the Arctic and the Antarctic (followed by Canada, the United Kingdom, Germany, Norway, and Russia). IPY projects thus benefited from substantial U.S. participation and leadership. IPY's focus on international partnerships motivated many researchers to expand their collaborations with scientists with similar interests in other nations. For an IPY project to receive ICSU-WMO endorsement, teams were required to include members from several nations. This cardinal feature gave IPY efforts an internationally recognized imprimatur and encouraged the leveraging of multinational infrastructure and intellectual assets, thus increasing the impact and capability of the project teams.

Measures to fuse international research teams into larger groupings addressing closely related topics were adopted by the IPY Joint Committee (JC) and International Programme Office (IPO) early in the planning process. In 2005 the IPO worked with the JC in a transparent process to collect and review more than 1,000 Expressions of Intent (short descriptions of proposed projects) and urged contributors to partner with other teams for larger or coordinated ventures. These Expressions of Intent eventually became 422 full proposals with broader focus and larger team size, and of these, 228 were recommended for implementation as "endorsed international projects" (Krupnik et al., 2011[3]), including a number of large-scale international initiatives engaging scientists from multiple nations. Many projects featured particularly strong U.S. involvement, among them the Integrated Arctic Ocean Observing System (iAOOS[4]); Polar Study using Aircraft, Remote Sensing, Surface Measurements and

[2] Aksnes and Hessen (2009) examine the period 1981-2007; there is an updated publication in preparation that examines more recent data including the IPY time period; the relative contributions of the countries listed here are unchanged (Dag Aksnes, Nordic Institute for Studies in Innovation, Research, and Education, personal communication).

[3] *Understanding Earth's Polar Challenges: International Polar Year 2007-2008* (Krupnik et al., 2011), is a summary report of IPY activities written by the JC. It includes coverage of IPY history as well as a broad overview of international contributions during IPY 2007-2008.

[4] *http://aosb.arcticportal.org/programs.html.*

Models, of Climate, Chemistry, Aerosols, and Transport (POLARCAT[5]); GEOTRACES[6]; Census of Antarctic Marine Life (CAML[7]); Circumpolar Biodiversity Monitoring Program (CBMP[8]); Arctic Human Health Initiative (AHHI[9]); Antarctica's Gamburtsev Province Project (AGAP[10]); and Norwegian-U.S. Traverse of East Antarctica.[11] (See Chapter 3 for more information.) In addition, the United States funded IPY-relevant projects from national IPY solicitations that were not submitted to the JC vetting process.

Later in this chapter, a section on Diversity in the Polar Research Community highlights the sharp contrast in the participation of women since the IGY in 1957-1958: the number of female principal investigators has increased in the United States over the past 10 years and, whereas U.S. women were virtually absent from the IGY effort, they held strong leadership positions in all phases of the recent IPY.

IPY was tremendously important for engaging with Arctic communities, including capacity building in areas with little previous experience in polar research (Krupnik et al., 2011). One major result was a sea change in the degree of engagement and active participation of polar residents and indigenous peoples (see below and Chapter 5). Arctic residents participated in many IPY events, resulting in a sharing of observations and interpretations with scientists who often study the poles remotely or visit briefly during the summer. These expanded avenues of collegial interaction during IPY offered both polar researchers and Arctic residents more new and varied opportunities to enhance their understanding of the Arctic.

In addition to connections among scientists engaged in research, IPY outreach activities and special sessions at scientific meetings fostered communication and association. New levels of interaction inevitably developed as a result of scientists presenting their stories and results to audiences at numerous scientific, education, and outreach meetings over the almost 8-year-long period of IPY planning and implementation (2003-2010). This was certainly the case at large-scale international gatherings such as the AGU and European Geosciences Union meetings, the Scientific Committee on Antarctic Research and the International Arctic Science Committee "open science" conferences, and the two major IPY conferences in 2008 in St. Petersburg and 2010 in Oslo, each of which engaged several thousand participants from many nations.

## TRAINING YOUNG SCIENTISTS

Most of the science funded by the United States for IPY efforts included support for graduate or undergraduate students, who worked closely with students and faculty from other nations and became central players in both national and international projects. They were also involved in outreach, learning the importance of communicating science to a broader audience. Many of the projects focused on the Arctic created direct connections between the students and residents of the region, affording the students a true appreciation of the capabilities and experiences of residents and their adaptations to climate change.

An example of an IPY activity that facilitated interdisciplinary relationships among the new generation of polar researchers was the NSF-funded Next Generation Polar Research Symposium in 2008 (Weiler et al., 2008). This symposium enabled a diverse group of scientists new to the polar community to interact with both active and retired polar scientists and thus provided a new generation with a common sense of history and research connections for the future.

Another avenue of student participation was the University of the Arctic,[12] a network of higher-education institutions and organizations established in 2001 to promote knowledge, research, and sustainability in the North. The network consists of over 130 member organizations across 8 nations and offers Arctic-focused courses and joint programs, often in partnership with indigenous peoples, and its membership and student enrollment have steadily increased since its inception. Its importance was evident during IPY as it helped to coordinate education and outreach activities associated with international research projects.

---

[5] *www.polarcat.no/.*

[6] *www.geotraces.org/.*

[7] *www.caml.aq/.*

[8] *http://caff.is/monitoring.*

[9] *www.arctichealth.org/ahhi/.*

[10] *www.ldeo.columbia.edu/res/pi/gambit/.*

[11] *http://traverse.npolar.no/introduction.*

[12] *www.uarctic.org.*

An outstanding success of IPY was the establishment in 2006 of the Association of Polar Early Career Scientists (APECS[13]). Sparked by a small core of enthusiastic and imaginative young polar scientists and a supportive IPO, the group made the most of the quick and effective social networking tools that are especially familiar to the young in this era of global electronic communications. APECS has already become "the preeminent international organization for polar researchers at the beginning or early stages of their careers."[14] This self-started activity so fully addressed the IPY objective of engaging young scientists in polar research that APECS developed its own early-career program and integrated it into the overall suite of IPY activities.

APECS achievements during IPY included the creation of an international and interdisciplinary network for early career polar scientists to share ideas, develop new research directions, and form collaborations; promotion of education and outreach as integral components of polar research to stimulate future generations of polar researchers; and arrangement of opportunities for professional career development through webinars, workshops, and session leadership at symposia. Indeed, the participation of APECS members among the speakers, planners, and session cochairs of major IPY-related meetings became essential.

APECS has grown at an extraordinary rate, to a total of 2,652 members from 45 countries as of July 2011 (Figure 2.2); nearly 20 percent (499) of APECS members are in the United States. APECS receives increasing support and endorsement from many international organizations. Importantly, the association has not only survived the rapidly changing careers of its leadership but thrived as creativity and energy are replenished, making it one of the most vibrant legacies of IPY. APECS is a model to inspire youth to consider the exciting and rewarding potential of a polar research career.

## INCREASING DIVERSITY

The polar research community has never been particularly diverse—50 years ago, planning for U.S. efforts in the IGY was carried out by an all-male committee. In contrast, women had strong leadership roles in planning U.S. involvement in IPY 2007-2008. There was a female director of the Polar Research Board (PRB), a female chair of the PRB during most of the IPY years, and a female chair of the U.S. National Committee for IPY Vision (NRC, 2004).

Similarly, a significant difference between IPY 2007-2008 and its predecessors was the participation of women in project leadership and participation—an increase from almost no female leads during IGY to around one-fourth during IPY. As shown in Table 2.1, during the 10 years from 1999 to 2009, the number of female project leaders increased by 10 percent, with an overall increase of 6 percent during that period in women among principal and coprincipal investigators. Despite recent developments, women and indigenous peoples in particular remain underrepresented.[15]

The research community remains far less diverse in terms of race and ethnicity, but IPY offered an excellent opportunity to increase diversity through the involvement of graduate and undergraduate students in the science programs (Figure 2.3). For example, the Dartmouth College NSF-funded IGERT (Interdisciplinary

TABLE 2.1 NSF/OPP Grant Recipients by Gender, 1997-1999 and 2007-2009

| | PI | | | Co-PI | | | Total | |
|---|---|---|---|---|---|---|---|---|
| | Total | Male | Female | Total | Male | Female | Male | Female |
| 1997-99 | 742 | 84% | 16% | 330 | 76% | 24% | 82% | 18% |
| 2007-09 | 1051 | 74% | 26% | 521 | 79% | 21% | 76% | 24% |

NOTE: To assess the participation of women in IPY, projects supported by the NSF Office of Polar Programs (OPP) were targeted as a representative sample. Note that these may include education and outreach as well as research projects supported by OPP. Grant recipients were categorized by gender in both 1997-1999 ($n$=1072) and 2007-2009 ($n$=1572), by denoting gender-obvious names (e.g., "John" = male; "Clara" = female) and researching and clarifying the remaining names. Genders were thus determined for more than 99 percent of principal investigator (PI) names. SOURCE: Data from *http://www.nsf.gov*.

[13] *www.apecs.is/*.

[14] Jenny Baeseman, APECS Director, personal communication, 2011.

[15] *www.ldeo.columbia.edu/res/pi/polar_workshop/strategies/communities.html*.

FIGURE 2.3 U.S. Geological Survey (USGS) field engineer Beth Burthon (left) and graduate student Adrienne Block (right) from Lamont Doherty Earth Observatory analyze data in the field as part of Antarctica's Gamburtsev Province project. SOURCE: Robin Bell.

Graduate Education, Research, and Training) program on Polar Environmental Change, which started during IPY in partnership with Greenlander Aqqaluk Lynge and the Inuit Circumpolar Council of Greenland, was successful in recruiting minorities to this interdisciplinary PhD program; among others, a female Native American biologist and black female electrical engineer are completing their PhD degrees. Another example is the Research and Educational Opportunities in Antarctica for Minorities (IPY-ROAM) program,[16] in which university students and high school teachers travel to Antarctica and learn firsthand about research in the field.

Some IPY outreach activities specifically targeted certain audiences to deliver the message that polar research is an equal-opportunity career choice. Of particular note was the 2008 National Annual Conference of the Society for the Advancement of Chicanos and Native Americans in Science titled "International Polar Year: Global Change in Our Communities."[17] The meeting, which resulted from the determined efforts of the IPO, enabled an important dialogue among Native Elders, students, and polar scientists from many disciplines about the health of the poles, polar peoples, global climate concerns, and ways to make positive contributions to the sustainability of the planet.

## ENGAGING POLAR RESIDENTS AND BUILDING COMMUNITY CAPACITY

IPY represented a sea change in bringing Arctic residents and indigenous peoples into polar research. Both constituencies were valuable contributing members to IPY activities by virtue of their expert knowledge of local environments and their involvement in IPY data collection, local observations, and education and outreach activities.

Many IPY projects encouraged this trend through grants that fostered collaboration with indigenous people and through student training about the importance of local communities and of good communication with people living in the Arctic. This was a particularly notable aspect of IPY as Arctic residents had little if any role in the earlier IGY/IPYs, whereas in IPY 2007-2008 they launched or led four projects and were active in more than 20 others (Gofman and Dickson, 2011). Also of particular importance was the participation of indigenous experts—elders, hunters, reindeer herders, and others—as long-time environmental monitors and researchers in several IPY projects, such as Sea Ice Knowledge and Use (Krupnik et al., 2010b), the Bering Sea Sub-Network (BSSN[18]), EALÁT (Reindeer Herders Vulnerability Network Study[19]), and others.

IPY promoted the practice of returning usable data to communities (see section in Chapter 5 on "Providing Critical Information to Users and Decision Makers"). Furthermore, for the first time IPY data were collected and disseminated in indigenous languages: Inuit (Inuktitut, Kalaallit, Iñupiaq, Yup'ik, and Yupik), Gwich'in, Sámi, Chukchi, Sakha, and Nenents. Also for the first time, IPY activities documented and supported indigenous languages and knowledge in the Arctic regions, including endangered Native languages (in Alaska), knowledge of marine animals, terrestrial animals, and sea ice (funded by NSF and the National Park Service [NPS]).

These successes were despite the fact that few if any indigenous representatives were active on the IPY governing bodies at either the international or national level (Krupnik et al., 2011). Also, the impact of IPY was very uneven across polar communities. Some (e.g., Barrow, Togiak, and Gambell in Alaska; Igloolik,

[16] *http://ipyroam.utep.edu/.*

[17] *http://sacnas.org/about/stories/tek.*

[18] *http://www.bssn.net/.*

[19] *http://www.arcticportal.org/en/icr/ealat.*

Clyde River, and Iqaluit in Canada; and Kautokeino in Norway) were actively engaged and informed, whereas many more had sporadic or ad hoc access to IPY information and resources. This is a valuable lesson for future planning.

U.S. scientists, together with their colleagues from Canada, Norway, Russia, Sweden, and Greenland, were at the forefront of partnerships with polar residents, as almost a dozen U.S. IPY projects funded by NSF, the National Oceanic and Atmospheric Administration (NOAA), USGS, NPS, and other agencies engaged northern residents and indigenous people in data collection and other research activities. Research grants to nonacademic groups further expanded and diversified the body of polar researchers during IPY.[20] Such grants were fairly unusual for the NSF OPP, and the committee applauds the OPP for thus enabling organizations that had longstanding relationships with local communities to conduct some IPY activities, including targeted workshops, websites, webinars, exhibits, popular books, performance art, teacher training, and public programs. Several activities targeted families and children with the aim of exciting future generations in polar research and showing parents what was being learned and why it mattered. Examples of these programs include "Beyond Penguins and Polar Bears: Integrating Literacy and the IPY in the K-5 Classroom" at Ohio State University, and "Penguins Teaching the Science of Climate Change" by Harvey Associates.[21]

In addition to special events for different audiences (e.g., the general public, government officials, educators, schoolchildren, and nongovernmental organizations), IPY featured broadscale public activities, starting from its official opening in March 2007. Most notable were seven online "International Polar Days"[22] (some of which actually lasted a full week) that featured IPY activities about sea ice (September 2007), ice sheets (December 2007), changing Earth (March 2008), land and life (June 2008), people (September 2008), outer space (December 2008), and polar oceans and marine life (March 2009). Two polar weeks in October 2009 and March 2010 focused on community building.

In the United States, "Polar Weekend" science fairs provided a forum for polar researchers, educators, and performers to engage with a public audience through hands-on displays and presentations. There were three such events in New York at the American Museum of Natural History (2007, 2008, 2009, with more than 12,000 visitors) and two in Baltimore at the Maryland Science Center (2009, 2010). Each fair involved about 100 presenters/volunteers representing about 30 institutions. Recurring Polar Weekends were also held in Seattle, Washington, and Fairbanks, Alaska; and sporadically at other locations including Kansas, Michigan, and Illinois. Surveys indicated that the public strongly valued these face-to-face interdisciplinary programs. By engaging participants in developing presentations, activities, and resources for the general public, these fairs also built the capacity of polar researchers to become active and articulate spokespeople during IPY and beyond. A large proportion of the participants (46 percent) indicated that "communicating with the public is now something I consider part of my career."[23]

The Exploratorium, a museum in San Francisco, put the power of the camera and written word into the hands of researchers in producing "Ice Stories," an online resource that engaged the public in the adventures of polar researchers in their pursuit of scientific discoveries. Two years of webcasts from the Arctic and Antarctic provided an "up close and personal" look intended to help the public relate to science and scientists.

Of the myriad IPY outreach activities, the jointly funded NSF and National Aeronautics and Space Administration (NASA) Polar-Palooza[24] project had a particularly high public profile. Its cadre of 34 polar scientists and Arctic spokespeople performed at 24 museums and science centers across the United States before extending the group's reach to venues in Argentina, Australia, Brazil, China, Norway, and Russia, and the podcasts and online activities and materials were used in many more countries.

Polar-Palooza's "Stories from a Changing Planet" consisted of a professionally produced stage show with exciting music and stunning photography as a backdrop against which scientists and Arctic residents presented

---

[20] *http://www.nsf.gov/od/opp/ipy/awds_lists/final_awrds_lists/ipy_awrds_rev02032011.pdf.*

[21] *http://www.nsf.gov/od/opp/ipy/awds_lists/2010_awds/ehr_awds.jsp.*

[22] *http://ipy.arcticportal.org/feature/item/1113.*

[23] Stephanie Pfirman, Barnard College, personal communication, 2011.

[24] *http://passporttoknowledge.com/polar-palooza.*

important scientific data and compelling research. These efforts complemented academic programs to train students to work with local communities. Eventually, the combination of these two efforts should continue to benefit both research and local polar communities, promoting not only more rigorous research methods but also greater communication, trust, and collaboration between the two communities.

## COMMUNICATING WITH THE PUBLIC

Extensive outreach and communication of science results to the public were a priority objective of U.S. IPY activities (NRC, 2004). As a result, each project had to include outreach and education activities as a condition of endorsement. In addition, 57 of the international IPY projects (out of more than 228 total) specifically focused on communicating IPY science to the broader science community, students, educators, policymakers, and general public in a variety of personal experiences (Figure 2.4). The show enabled face-to-face discussions between Arctic residents experiencing the undeniable impacts of climate change and midlatitude citizens for whom the concept of climate change was still difficult to comprehend. The opportunity to meet a polar researcher "in the flesh," to hold the tools used, and to see a real ice core were all powerful methods of engagement.[25] The stage shows were often accompanied by additional pole-related activities in local museums and visits to schools to further the engagement of polar researchers with public audiences, educators, and schoolchildren.

Formal outreach activities included organizer and participant evaluations to quantify their effectiveness. These assessments showed that many outreach activities were successful in informing their audiences of the seriousness of observed changes in polar and global climate and of the role of polar research in supporting those conclusions (Perry and Gyllenhaal, 2010). According to the assessment report for Polar-Palooza,

> There were strong indications that the Polar-Palooza model of using real scientists and Alaska Natives as presenters worked very well for most Polar-Palooza audience members, whether they attended and participated in the presentations, the educator workshops, and/or the outreach activities and events. Under the guidance of Polar-Palooza staff working with them both in advance and "'on-the-fly,'" the . . . presentations and the multimedia framework of graphics and high-definition video formed an engaging and coherent whole for most respondents. (Perry and Gyllenhaal, 2010)

Although the IPY community receives admiration and acclamation for its collective accomplishments in education and outreach, it has been suggested that more tangible means for professional recognition need to be developed (Salmon et al., 2011). What the research community learned, in return for their efforts, is that the human dimension is essential not only in the conduct of science but also in its communication. Participant feedback repeatedly indicated that the adventure of the endeavors, the importance of the science, and the thrill of discovery were essential to engaging the public.

## PROVIDING RESOURCES FOR TEACHERS

The effort of informing teachers and other educators about polar science can have long-term benefits, and IPY included a number of programs that engaged teachers. IPY scientists worked with the National Science Teachers Association (NSTA) to reach science teachers around the world. NSTA coordinated several symposia (face-to-face workshops) at their area and national conferences, and many teachers said that direct access to scientists was one of the most exciting and valuable features of the programs. NSTA also worked with representatives of multiple federal agencies (NASA, NOAA, and NSF) and produced effective web seminars (webinars) using the association's expertise, access to a network of 400,000 science teachers, and portal (the NSTA Learning Center). Of enduring benefit are dozens of webinar archives and podcasts that are available to all teachers free of charge and on demand via the Learning Center—"evergreen" productions that teachers consume on a regular basis. Science teachers can use these resources for years to come to inform students, illuminate the human dimension of polar research (e.g., by talking about science careers), and hence increase the human capacity of polar research.

[25] *http://ipy.arcticportal.org/feature/item/1113.*

FIGURE 2.4 Photos from Polar-Palooza events. Top left: Mary Albert explains carbon dioxide and temperature data from the Vostok ice core. Middle left: Audience at a Polar-Palooza Event. Bottom left: Graduate student Atsu Muto describes the archiving of ancient atmospheres in bubbles in the polar ice sheets to local citizens. Right: Polar biologist Dr. Michael Castellini assists a ìmid-latitude penguinî in answering questions from the audience at the Polar-Palooza event in Fort Worth, Texas. SOURCES: Geoffrey Haines-Stiles and Polar-Palooza website (*http://passport toknowledge.com/polar-palooza*).

Prepared teaching lesson plans that explicitly address national science standards and other curriculum requirements were made available to teachers. Many of the larger IPY science programs included an education and outreach office to create materials for teachers. One such program is Antarctic Geological Drilling (ANDRILL), which created a variety of programs and activities for teachers during the IPY years as well as an education event for teachers at the IPY Oslo Science Conference in 2010.[26]

A science education initiative that reached large numbers of teachers and their students in the United States and other countries was Monitoring Seasons through Global Learning Communities, also called GLOBE (Global Learning and Observations to

[26] *http://www.andrill.org/education.*

Benefit the Environment) Seasons and Biomes.[27] This inquiry-based project monitors seasons, and specifically their interannual variability, to increase students' understanding of the Earth system (its focus has progressed from the tundra and taiga biomes to temperate, tropical, and subtropical forests, grasslands, savannahs, and shrublands). It links students, teachers, scientists, citizens, and local experts in more than 50 countries. Seasons and Biomes has conducted 32 professional development workshops for 600 educators and scientists, and those who have been trained conducted an additional 23 workshops for 436 teachers and 20 preservice teachers, reaching more than 1,000 educators and an estimated total of 20,000 students from all over the world.

PolarTREC (Teachers and Researchers Exploring and Collaborating) is an NSF-sponsored program that pairs K-12 public school teachers with scientists working in the field in the Arctic and Antarctic. The teachers participate in a science field team in the Arctic or Antarctic and relay their experiences and adventures to students at home and around the world through blogs and webinars. Once the field season ends, the researchers traveled to the teachers' schools to make presentations about the science and talk with the students.

During IPY, PolarTREC enabled 48 teachers (from elementary through high school) to engage approximately 5,000 students in polar research activities.[28] Students and teachers alike reported an increase in their understanding of the polar regions and of scientific processes and practices after the PolarTREC experience. Similarly, scientists that participated in the program found that they were able to better communicate science to a K-12 audience. The PolarTREC website[29] serves as an archive of webinars and other resources for classroom activities.

[27] *http://classic.ipy.org/development/eoi/details.php?id=278.*

[28] *www.polartrec.com/expeditions/prehistoric-human-response-to-climate-change-2010.*

[29] *www.polartrec.com.*

## MAINTAINING AND INCREASING HUMAN CAPACITY

IPY invested heavily in the development of younger scholars, partnerships with polar residents, training of school teachers and students, and outreach to the general public, all with the goal of increasing the human capacity of polar science. There is early evidence that these "people-focused" IPY activities were positively received and will support further efforts to inform and engage the public in polar issues. Continuing efforts may build on the fact that polar research yields important results that are of interest to the public as a useful source of information about large-scale climate changes and their societal impacts. Whether in the physical or social sciences, human health or public policy, there is a lot of information of use to diverse audiences. The need for experts in all these fields to investigate, understand, and respond to change is imperative.

Follow-up studies are the only certain means to track the staying power of the many IPY outreach efforts. Decadal or half-decadal surveys of indicators—such as the number of active researchers in various disciplines, number of polar residents and communities partnering with scientists, and public knowledge of polar-related issues—would be very useful to the long-term assessment of IPY.

## CONCLUSIONS

The committee identified the following positive outcomes of IPY efforts designed to engage the public and build human capacity for polar research:

- The emergence of the young scientists' peer network APECS, which provided a means for early career scientists in various countries to share ideas, develop new research directions, and form collaborations;
- The University of the Arctic, a network of higher-education institutions and organizations, helped to coordinate education and outreach activities internationally;
- Major increases in the numbers of researchers, women, and polar residents actively involved in polar research;
- A significant expansion of collegial links between experienced and new polar researchers, which

in turn benefited the creation of complex international projects;

- A new era of extensive and effective outreach and educational activities, in which polar science garnered increased attention and was enthusiastically received;
- The engagement with and inspiration of teachers by polar scientists through the extensive network of teachers associated with the NSTA, as well as the legacy of webinars and classroom materials that met national standards for education and provided a long-term repository of resources and activities; and
- A new impetus to engage educators in teaching polar sciences, and to use polar science to draw students into science more broadly, matched by an increased level of activity in the polar research community to meet the needs of teachers and their students.

The education and outreach efforts during IPY raised the bar for the quality of scientist interactions with teachers, students, and the public in ways that increase public understanding of science. Continued funding for strong and effective outreach activities such as those described in this section is critical to continued public engagement in and understanding of polar science.

# 3

# Scientific Advances and Discoveries

In the golden age of polar exploration at the beginning of the 20th century—when explorers like Frederick Cook, Robert Peary, Roald Amundsen, Robert Scott, and Ernest Shackleton led expeditions to the North and South Poles—science discovery was driven by the desire to fill in blank spaces on maps. Now it is driven by the desire to learn about different kinds of unknowns, such as the consequences of changing climate, ecosystems that exist on the underside of the ice, changing patterns of sea ice, and mechanisms of ice sheet flow. International Polar Year 2007-2008 witnessed a host of discoveries in a wide variety of scientific fields—from how Earth's climate has changed in the past to how it may change in the future, from understanding what goes on in the depths of the ocean to understanding the weather out in space, and from learning about the impacts of climate change on marine ecosystems to their implications for human societies.

Many IPY discoveries relate to how quickly Earth is warming (Box 3.1). Contemporary change detection is difficult for some polar climate processes and their global linkages because the frequencies of natural change may be longer than the three decades of modern observations. It is clear, however, that global weather and climate patterns are interconnected yet spatially variable. The polar regions play key roles in this global system, in part because of the interactions of land and ice masses with ocean currents and atmospheric circulation patterns. IPY enabled scientists around the world to join forces, using new tools and bridging frontiers, in seeking information to expand knowledge of these patterns and Earth's linked systems. Specifically, scientific results from IPY projects are shedding light on changes in the environment, including climatic responses and human-environmental dynamics.

## POLAR ICE SHEET SCIENCE AND SUBGLACIAL SYSTEMS

During IPY, numerous international teams addressed scientific issues from on, within, and below the world's two massive ice sheets, in Greenland and Antarctica (Figure 3.1). Together these ice masses contain enough water to raise global sea level by 70 m if they melted. These vast frozen expanses exist because ice loss through melting, ablation, and the calving of icebergs has been balanced or exceeded in the past by winter snowfall. But in the years leading up to and during IPY, vivid images showed ice calving along the perimeters of both Greenland and Antarctica and the rapid disintegration of vast ice shelves along the Antarctic Peninsula. Although scientists are still evaluating relationships between ice mass loss and warming of the atmosphere and oceans, it appears that continued warming will likely cause more mass loss.

The sensitivity of the Greenland and Antarctic ice sheets to climate change and their major roles in modulating sea level make their condition relevant to society. Today millions of people live along low-lying coastlines within a meter of sea level. Future sea level rise is a concern for both maritime societies and extensive seaside commercial and military infrastructure throughout the globe (see also Chapter 5). Numerous satellite data enabled international teams to make important

## BOX 3.1
## Climate Change and Polar Amplification

The carbon dioxide content of the atmosphere has been steadily increasing from its preindustrial value of ~280 parts per million (ppm) (Figure) (IPCC, 2007a), and the rate of increase has significantly intensified in the past 60 years, from 0.53 ppm per year in 1957 (when measurements were started) to nearly 2 ppm per year in 2008. If this trend continues, by 2057 (perhaps the time of the next IPY) carbon dioxide values may approach 500 ppm. Furthermore, there is increasing documentation that in the geologic past, continental ice sheets grew only when atmospheric carbon dioxide values were low and retreated when values were high. Such trends are in agreement with climate change projections based on increasingly sophisticated computer simulations of climatic change.

Climate changes and their impacts have considerable spatial variability due to Earth's interlinked and moving atmospheric and oceanic circulation patterns. Although warming of the planet as a consequence of anthropogenic carbon dioxide is happening nearly everywhere (IPCC, 2007a), it is accentuated in the high latitudes due to polar amplification caused by strongly positive snow and ice feedbacks (Serreze et al., 2009; Manabe and Wetherald, 1975). Concern about this polar amplification drove much of the research carried out during IPY, and the complex nature of Earth systems and changes will require significant and continuing investigation to foster understanding and action toward sustainability.

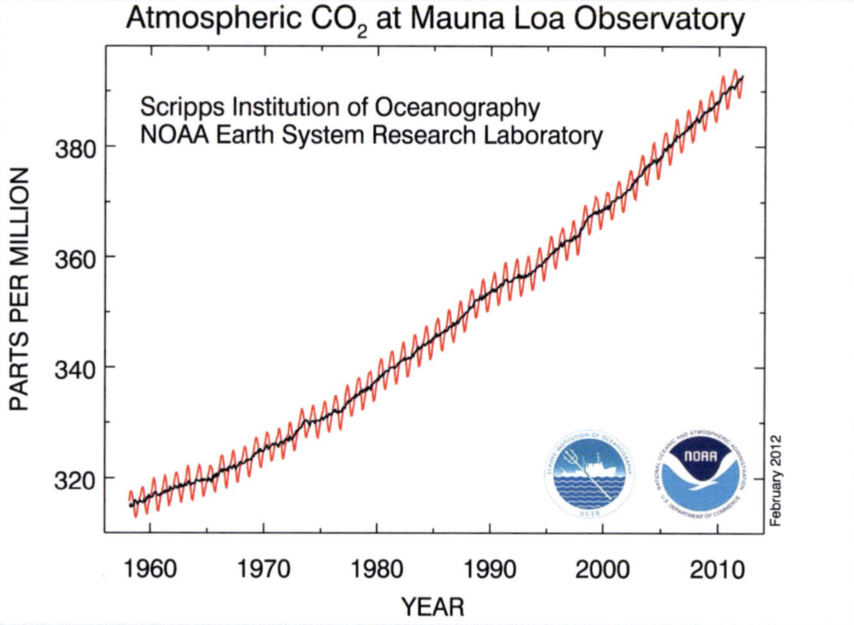


FIGURE The carbon dioxide ($CO_2$) data (red curve) measured on Mauna Loa constitute the longest record of direct measurements of $CO_2$ in the atmosphere. The black curve represents the seasonally corrected data. SOURCES: NOAA Earth System Research Laboratory and Scripps Institution of Oceanography.

discoveries of large changes occurring in both ice sheets. Taken together, these assessments showed that the pace of ice sheet mass loss has been increasing since the end of the last century, accelerating sea level rise. As a result of research coming out of IPY, projections for the future show an accelerating trend for sea level rise by 2100, with model predictions ranging from 20 to 180 cm (Figure 3.2). The upward trend in sea level rise is primarily the result of melting of glaciers and small ice caps, and the thermal expansion of seawater due to ocean warming. The former accounts for about 30 percent of the contribution to sea level rise (Nicholls and Cazenave, 2010).

Ice sheet mass changes can be estimated using data from Gravity Recovery and Climate Experiment (GRACE), a satellite designed to measure gravity variations, which in this case are created by regional mass redistributions within the ice sheets. IPY findings from the LEGOS[1] project using GRACE data provided evidence that the Greenland and Antarctic ice sheet contributions to sea level rise increased to 30 percent of the total sea level rise after 2003, compared to their smaller contribution of 15 percent of sea level change between 1993 and 2003 (IPCC, 2007a). Repeat observations of ice sheet elevations from the laser altimeter onboard ICESat-1 captured the detailed

[1] Laboratoire d'Etudes en Géophysique et Océanographie Spatiales; *www.legos.obs-mip.fr/*.

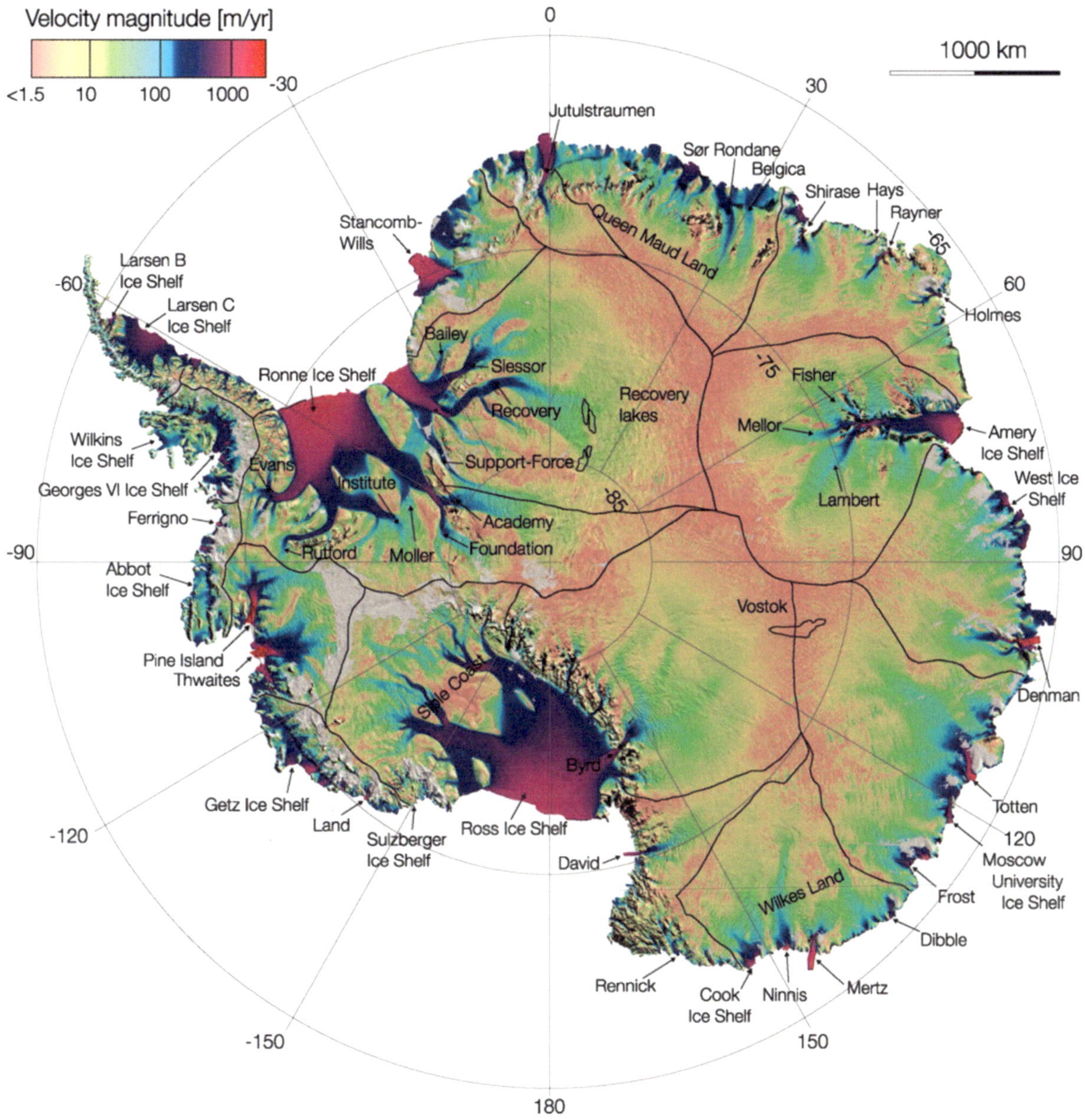

FIGURE 3.1 Maps of Arctic and Antarctic showing major ice sheets. Top: Antarctic ice velocity derived from ALOS PALSAR (Advanced Land Observing Satellite, Phased Array type L-band Synthetic Aperture Radar), Envisat ASAR (Advanced Synthetic Aperture Radar), RADARSAT-2, and ERS (Earth Remote Sensing)-1/2 satellite radar interferometry, color-coded on a logarithmic scale, and overlaid on a MODIS (Moderate Resolution Imaging Spectroradiometer) mosaic of Antarctica. Thick black lines delineate major ice divides. Thin black lines outline subglacial lakes. Thick black lines along the coast are interferometrically derived ice sheet grounding lines. Bottom: Satellite-based passive microwave images of the sea ice cover have provided a reliable tool for continuously monitoring changes in the extent of the Arctic ice cover since 1979. This visualization shows Arctic sea ice minimum area for 2010. SOURCE: Rignot et al., 2011 (top); NASA Scientific Visualization Studio (bottom).

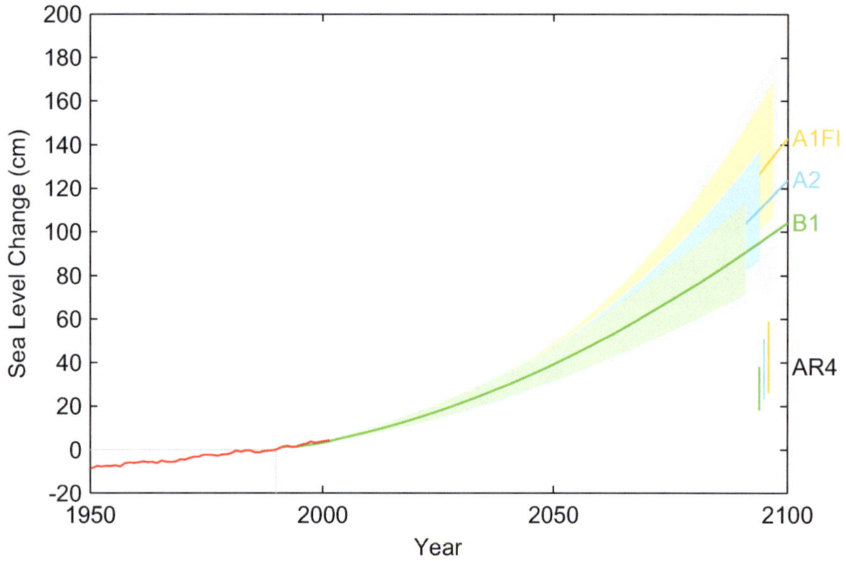


FIGURE 3.2 Projection of sea-level rise from 1990 to 2100, based on Intergovernmental Panel on Climate Change (IPCC) temperature projections for three different emission scenarios (labeled on right). The sea level range projected in the IPCC AR4 (IPCC, 2007b) for these scenarios is shown for comparison in the bars on the bottom right. Also shown is the observations-based annual global sea level data (Church and White, 2006) (red) including artificial reservoir correction (Chao et al., 2008). SOURCE: Vermeer and Rahmstorf, 2009.

pattern of changes across each ice sheet. This pattern of interior gain and marginal loss that had been seen in Greenland was detected across the Antarctic ice sheet with particularly large mass losses concentrated in the Amundsen Sea region (Pritchard et al., 2009).

The magnitude and pattern of ice mass loss directed many researchers' attention during IPY to the margins of the ice sheet, where the ocean melts the undersides of the floating fringes, and the base of the ice. Observations of accelerating ice flow following the collapse of Antarctic ice shelves confirmed the buttressing effect of ice shelves on ice flow and underscored the importance of ice shelves to ice sheet stability (Rignot et al., 2004; Scambos, 2011; Joughin et al., 2010). Initial oceanographic measurements made during IPY in the fjords of Greenland outlet glaciers (Figure 3.3) and beneath the Pine Island ice shelf in Antarctica confirmed the presence of warm waters expected to cause elevated ice melting (Jenkins et al., 2010; Rignot et al., 2010; Straneo et al., 2010).

The ramifications for ice sheet mass change of the discovery of a surprisingly active subglacial water system beneath both ice sheets were pursued vigorously during IPY. Abrupt draining of surface-melt lakes and associated fracture propagation and glacial movement were measured on the Greenland ice sheet (Das et al., 2008; Joughin et al., 2008). Discovery of actively filling and draining Antarctic subglacial lakes was followed by a comprehensive mapping using ICESat laser altimetry, resulting in the detection of 124 lakes actively filling or draining between 2003 and 2008 (Smith et al., 2010). The first case of a direct connection between subglacial lake drainage and a surge in an East Antarctic glacier was made using a combination of satellite altimetry and imagery (Stearns et al., 2008; Figure 3.4). Other lakes, detected with surface radar are water-rich but not currently water-filled (Langley et al., 2011).

New observations were made during IPY of the potential awakening of the East Antarctic ice sheet, a potentially powerful influence on global sea levels and the climate system. Surface mass balance in East Antarctica is a fundamental but poorly known component of the ice sheet mass balance (IPCC, 2007a). Using an interdisciplinary approach and latest technologies, the Norwegian-U.S. Scientific Traverse of East Antarctica IPY project aimed to improve understanding of surface mass balance and the drivers of climate variability in East Antarctica. Findings to date show that millennial-scale net accumulation rates from ice cores are generally lower than net snowfall estimated in previously published large-scale assessments (Anschutz et al., 2009). Discoveries of links between microstructure, accumulation rate, and satellite imagery provide the means of accounting for natural variability in 20th century accumulation trends and show that current climate models likely overestimate accumulation in East Antarctica (Albert et al., 2012; Scambos et al., 2011). Firn temperature measurements suggest a recent warming trend near the crest of the East Antarctic ice sheet but cooling or no change at a lower elevation

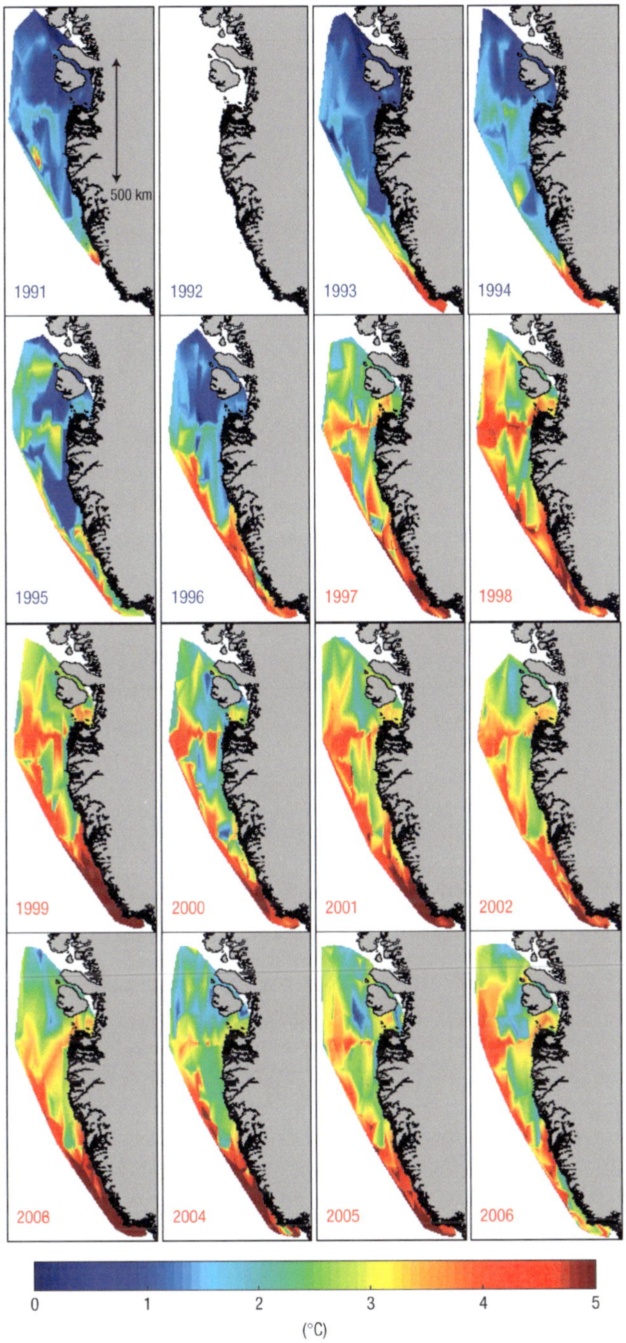


FIGURE 3.3 Warm waters located at the margins of an ice sheet can cause increased melting and thinning of the ice. This image indicates subsurface ocean temperatures over the west Greenland continental shelf showing the impact of warming fjord waters on the acceleration of ice flow. In 1997, a warm water pulse entered the region and coincided with rapid thinning and acceleration of an outlet glacier along the west coast of Greenland. SOURCE: Holland et al. (2008).

site (Muto et al., 2011). Aerosol depositions across all spatial and temporal scales across the continent show surprising similarity, supporting the importance of South American and other midlatitude source areas for dust, burning, and pollution aerosols (Bisiaux et al., 2012). Overall, our understanding of the East Antarctic Ice Sheet improved during IPY, and these observations set the baseline for continued monitoring in the future.

An entire mountain range under the Antarctic ice sheet was discovered during the 1957-1958 International Geophysical Year (IGY) but remained uninvestigated because of extreme inaccessibility until this IPY. The Antarctic Gamburtsev Province project used cutting-edge airborne radar to investigate the Gamburtsev Mountains which are completely covered by ice near the center of the East Antarctic Ice Sheet. The data reveal an Alps-like mountain range incised by fluvial river valleys in the south and truncated by the landward extension of the Lambert Rift to the north (Figure 3.5). Radar data reveal areas where hundreds of meters of ice have been frozen onto the bottom of the ice sheet, driving subglacial flow and ice sheet behavior in ways not captured in present models (Bell et al., 2011), as well as the first comprehensive view of the crustal architecture and uplift mechanisms for the Gamburtsevs (Ferraccioli et al., 2011).

IPY also created the opportunity for teachers, students, and laypersons, as well as scientists from many disciplines to access data to foster their own discoveries. The Landsat Image Mosaic of Antarctica (LIMA) IPY project produced the first-ever, true-color, high-resolution mosaic image of Antarctica using visible imagery from the years between 1999-2003 (Bindschadler et al., 2008). This tool, made available to the public,[2] enables teachers, students, and scientists to explore the continent from their desks. The early completion of this mosaic allowed polar researchers the opportunity to make discoveries during IPY using LIMA data. For example, using LIMA, Fretwell and Trathan (2009) identified previously unidentified penguin rookeries and determined other rookeries that have been abandoned.

Another one of the IPY 2007-2008 projects that had its beginnings during the IGY was the McCall Glacier Research Program (Weller et al., 2007). This

[2] *http://lima.nasa.gov.*

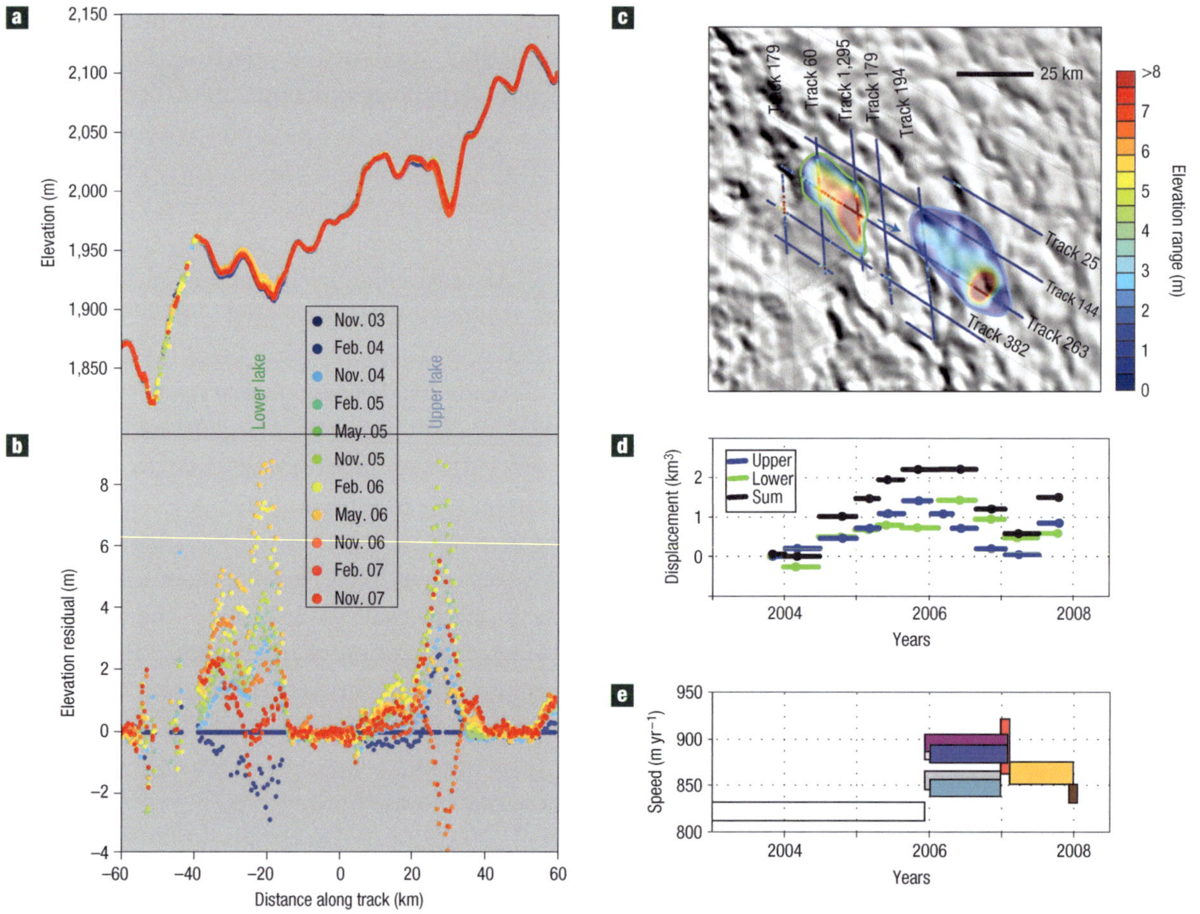


FIGURE 3.4 Subglacial systems (and the implications for ice sheet mass change) were actively studied during IPY. In fact, satellite altimetry and imagery were used to make a direct connection between subglacial lake drainage and changes in glacier dynamics. Notably, Stearns et al. (2008) show that surface elevation changes in the Antarctic Byrd Glacier reflect the filling and draining of the subglacial lakes below. (a) ICESat elevations for 11 passes between November 2003 and November 2007. (b) Elevation residuals for ICESat data after correction for topography. (c) Map of elevation ranges for 500-m sections of track, interpreted lake boundaries (green, blue outlines) and elevation ranges for gridded surface displacements. The arrow indicates the direction and orientation of the profiles in (a) and (b). (d) Estimated lake volume displacements for the downstream lake (green), the upstream lake (blue) and the two lakes together (black). The horizontal bars show time uncertainty in lake volumes. (e) Ice speed at the grounding line from 2003 to 2008. The horizontal bars indicate start and end dates for each pair of observations; the thickness of each bar represents its associated error. SOURCE: Stearns et al., 2008.

important glacier lies in the Brooks Range, near the northeast corner of Alaska and offers the longest history of research on any glacier in the U.S. Arctic. Studies revealed a long period of negative mass balance, with annual losses of −100 to −200 mm water equivalent during the period 1958-1971 (Nolan et al., 2005). The negative mass balance of the glacier persists today,[3] including accelerated thinning of ice volume and retreat of the ice margin, indicating that climate warming is exerting a long-term influence on glacial systems in this region.

## SEA ICE VULNERABILITY AND CONNECTIONS TO SOCIETY

Extensive research was conducted during IPY to understand the multifaceted role of sea ice—that is, ocean (salt) water that has frozen—in climate, ecologic, and socioeconomic systems. One critical element of sea ice that crosses many scientific disciplines is its influence on planetary albedo. Albedo, or reflectance,

[3] Matt Nolan, University of Alaska, Fairbanks, personal communication, 2011.

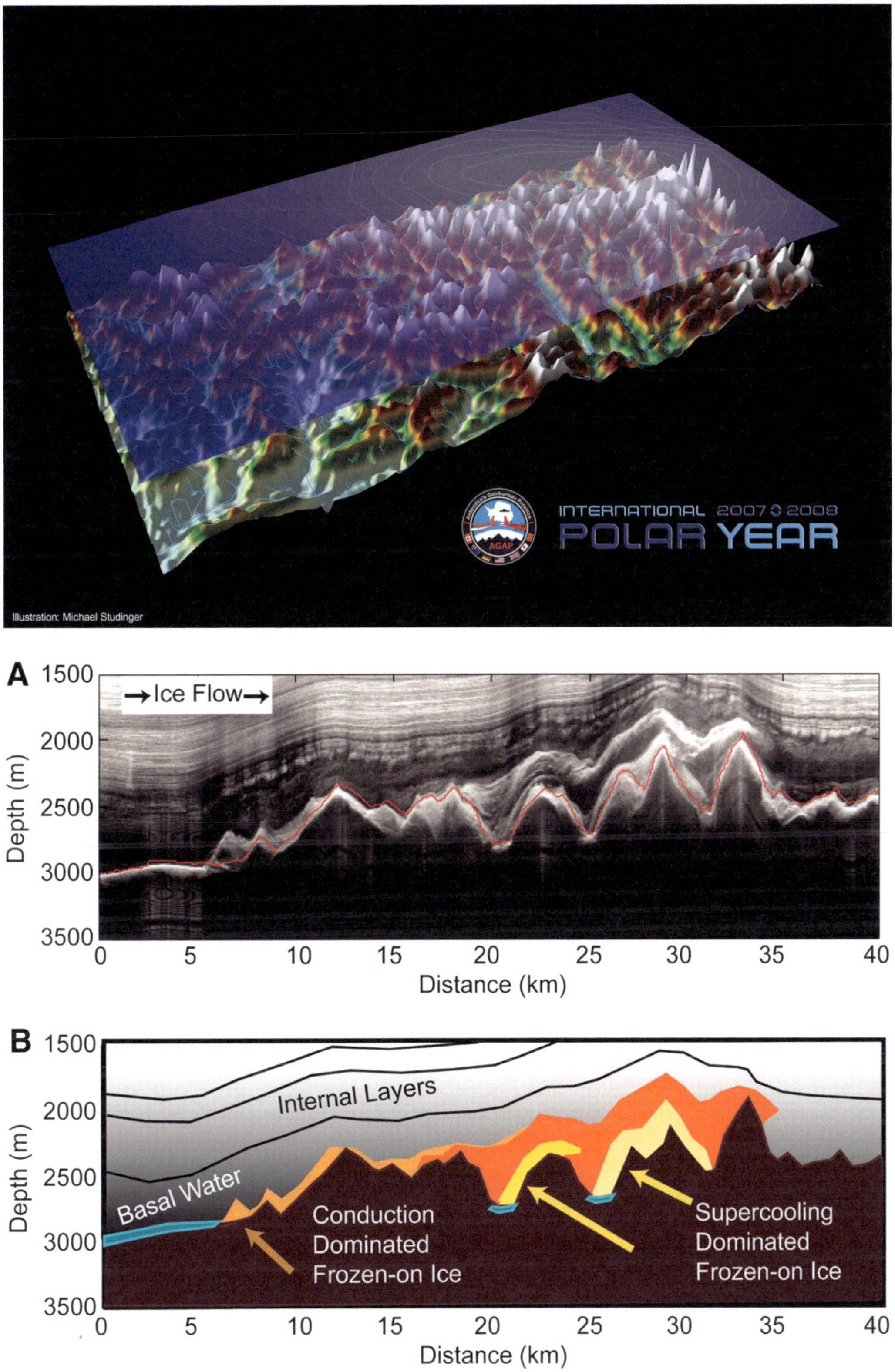


FIGURE 3.5 Top: Until IPY 2007-2008, an entire mountain range under the Antarctic ice sheet (discovered during the International Geophysical Year) remained uninvestigated because of inaccessibility. Cutting-edge airborne radar was used to investigate the Gamburtsev Mountains which are completely covered by ice near the center of the East Antarctic Ice Sheet. Bottom: (A) Radar data reveal areas where hundreds of meters of ice have been frozen onto the bottom of the ice sheet, driving subglacial flow and ice sheet behavior in ways not captured in present models. (B) A schematic of this process. SOURCES: Michael Studinger, NASA (top) and Bell et al., 2011 (bottom).

is a material property, dependent on the surface type. As the extent of the white sea ice and its associated snow cover decreases, more dark ocean surface water is exposed to the sun and warmed. This in turn melts more ice, further decreasing the extent of the ice, a positive feedback, and allows a significant increase in heating of the exposed ocean, leading to subsequent increases in the amounts of heat and moisture transferred from the ocean into the overlying atmosphere (Figure 3.6).

Because long-term studies of sea ice extent prior to 1950 were limited, and a bipolar record of the sea ice concentrations extending back to the late 19th century has only recently been attempted,[4] the science community largely relies on satellite-based passive microwave data collected since 1978. Instrumental time series of minimum ice extent leading up to 2006 showed that there had been an ~8 percent areal loss per decade in Arctic sea ice, with the strongest losses occurring in areas such as the Kara and Barents Seas. Thus IPY provided an opportunity to study the changing Arctic in a time of rapid change. During the first year of IPY investigations (2007), the minimum sea ice extent value showed a sharp additional decrease below most model projections from recent trends (Stroeve et al., 2007, 2008). This was not a trivial change; rather, Arctic sea ice area at the end of the 2007 summer was 27 percent lower than the previous record low observed in 2005 (Figure 3.7). Although in 2008-2009 there was a small recovery toward the earlier (pre-2006) trend-line, the 2010 value remained well below this trend. The sea ice minimum for late September 2011 was only slightly higher than in 2007.

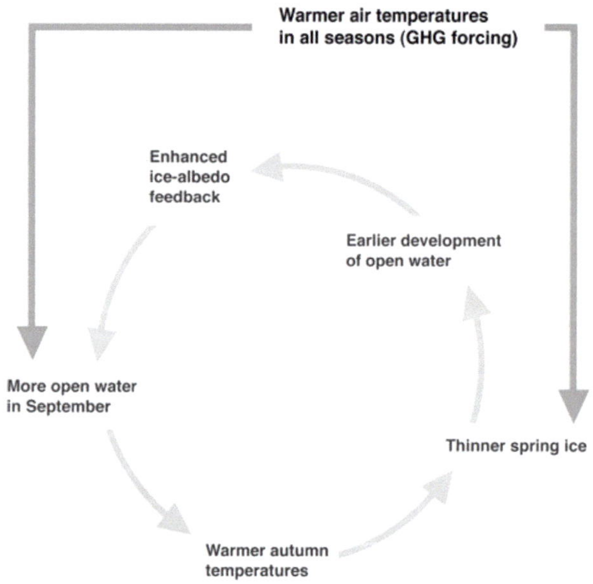



FIGURE 3.6 Cartoon of positive feedback leading to sea ice loss in the Arctic. SOURCE: Stroeve et al. 2011.

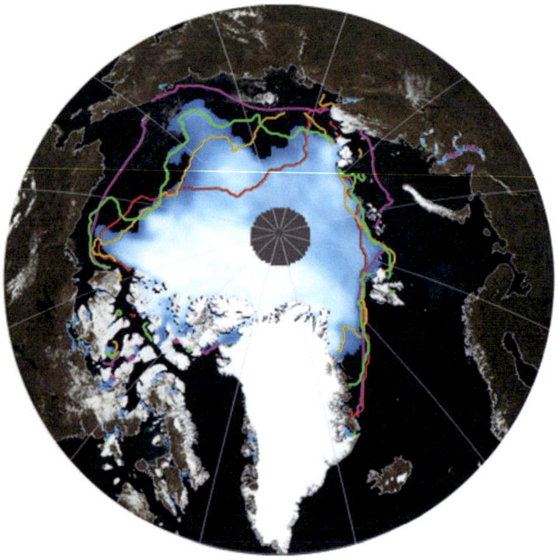

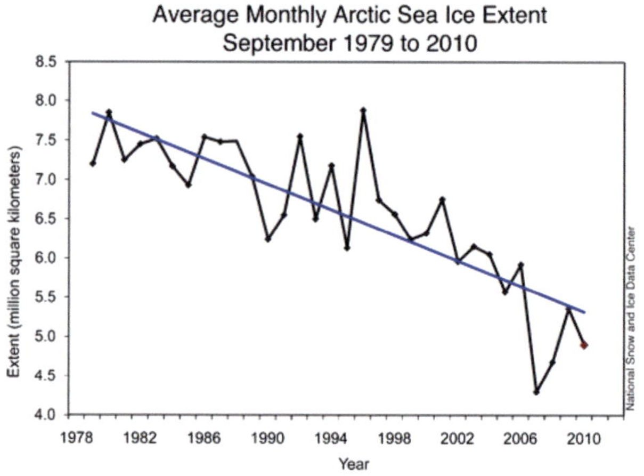



FIGURE 3.7 Instrumental records show a decrease in sea ice extent in the years leading up to IPY. As a result, IPY provided scientists an opportunity to study the Arctic in a time of rapid change. Top: Monthly sea ice concentration for September 2012. The red line marks the September 2007 extent, the orange line is the extent for September 2008, the green line the September 2009 extent, and the pink line is the climatological (1979-2000) monthly mean for September. Bottom: Time series of monthly averages of September sea ice extent with linear trend line showing decreasing trend over past three decades and sharp drop in 2007, the first year of IPY. SOURCE: Stroeve et al., 2011.

[4] *http://nsidc.org/data/docs/noaa/g00799_arctic_southern_sea_ice/index.html.*

In addition to changes in ice extent, upward-looking sonar data obtained via recent submarine cruises coupled with estimates based on measurements by NASA's ICESat laser altimeter indicate that sea ice thicknesses have also shown a substantial decrease of more than a meter from values obtained prior to 1990 (Kwok and Rothrock, 2009). Satellite tracking of ice trajectories (Fowler et al., 2003; Pfirman et al., 2004; Rigor and Wallace, 2004) indicated that the ice age—which correlates with thickness—was also decreasing (Figure 3.8). Although the minimum ice extent in 2011 was slightly larger than in 2007, the actual volume of ice present was dramatically less because of the consistent thinning trend due to the loss of thicker multiyear ice (Kwok and Rothrock, 2009; Maslanik et al., 2007). Many interpretations of these measurements were quickly published; some positing that 2007 represented a tipping point leading to an

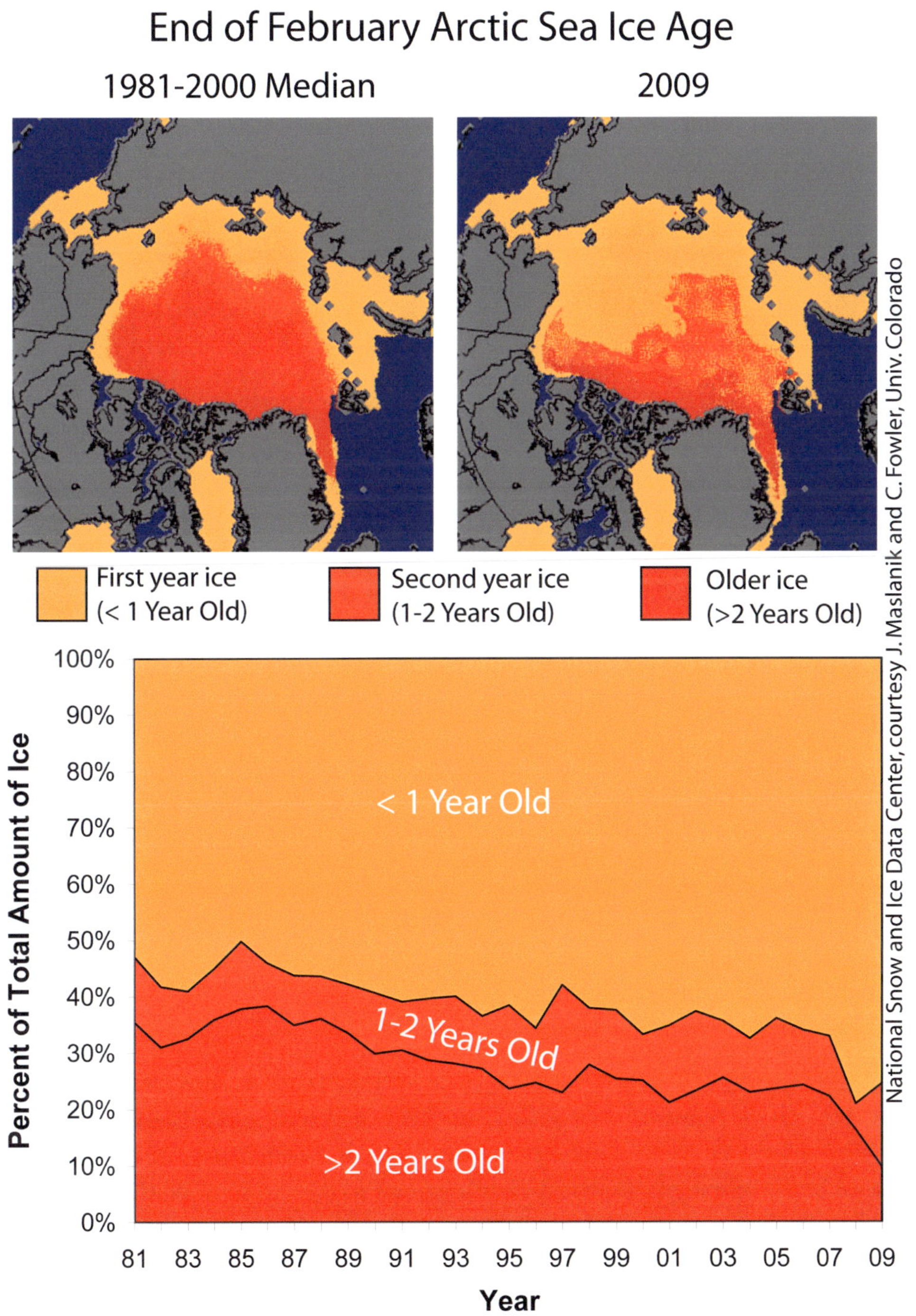


FIGURE 3.8 Satellite tracking provides information on Arctic sea ice age, which correlates with thickness. Upper panels show the spatial pattern of sea ice age at the end of February in 2009 (upper right) compared to the median age during the period 1981-2000 (upper left). The lower panel shows the relative percent composition of sea ice of different ages aggregated over the entire Arctic basin. It can be shown that sea ice cover in the Arctic is steadily decreasing both in extent and thickness, as indicated by the loss of multiyear ice. SOURCE: National Snow and Ice Data Center, courtesy J. Maslanik and C. Fowler, University of Colorado.

ice-free Arctic during Northern Hemisphere summers within the near future (Kwok and Cunningham, 2008; Stroeve et al., 2008). Replacing thick old ice floes that are currently being lost requires several years and is increasingly unlikely under current warming climatic conditions.

During periods of minimum sea ice extent, break-out events—such as the large volumes of ice swept out of the Arctic Basin during the fall of 2007—became more frequent (Stroeve et al., 2011). Current studies suggest that such events are the composite result of thermodyamic and dynamic processes, including the preconditioning of the ice by warmer than usual air and water temperatures and pressure patterns conducive to ice exiting the Arctic Basin via the Fram Strait (Perovich and Richter-Menge, 2009). In addition, studies of the ice-albedo feedback contribution to the observed warming of the upper layer of the Arctic Ocean show that although annual trends are small, cumulative effects are large, amounting to a 17 percent total increase by 2005. Although the time series of this contribution was fairly constant from 1979 to 1992, it then increased steadily from 200 $MJm^{-2}$ to about 400 $MJm^{-2}$. Furthermore overall negative trends in ice extent are also strongest in more northerly locations as would be expected if the ice-albedo effect was a significant contributor (Perovich et al., 2007).

Changes in Arctic sea ice extent are now becoming implicated with changing weather patterns at lower latitudes. During the last few years there has been a breakdown in the stable counterclockwise polar vortex wind pattern that during recent decades has characterized the Arctic. This wind acts to keep the far north cold and isolated from temperate regions farther south. As a result, cold air outbreaks to the south have become increasingly frequent, with the Arctic and sub-Arctic consistently exhibiting above-normal temperatures and less snowfall, while typically more temperate regions to the south have been subjected to heavy snows and frigid temperatures. Major economic disruptions occurred with these events in northern Europe, eastern North America, and eastern Asia. One contributing factor to these changes may be the decrease in both the extent and thickness of the Arctic sea ice cover (Serreze et al., 2007). At present it is uncertain exactly how recent changes in the Arctic are modifying sub-Arctic climate patterns such as the North Atlantic Oscillation (Overland, 2011), but this is an area of active research (Francis et al., 2009; Overland et al., 2007, 2008; Overland and Wang, 2010).

Sea ice conditions in the Antarctic differ from the Arctic in a number of ways. Most sea ice in the Southern Ocean melts each summer, so there are only a few regions with multiyear ice. The Antarctic Peninsula region is undergoing the largest positive temperature increase of any Southern Hemisphere location: +3.7 ± 1.6°C for the 20th century based on unweighted station data as compared to a global value estimate by the Intergovernmental Panel on Climate Change of +0.6 ± 0.2°C (Vaughan et al., 2003). These pronounced temperature changes have been associated with a noticeable decrease in sea ice extent (−5.4 percent per decade for the Amundsen and Bellingshausen Seas (Domack et al., 2003; Stammerjohn et al., 2008); and with spectacular retreats and disintegrations of several ice shelves, for example, Larsen, George VI, and Wilkins (Rignot et al., 2004; Scambos et al., 2004). In contrast, the sea ice cover around West Antarctica showed a slight increase in area with the Ross Sea showing the largest increase (+4.4 percent per decade; Cavalieri and Parkinson, 2008; Stammerjohn et al., 2008). When the Southern Ocean was examined as a whole, sea ice extent showed a slightly positive trend (+1.0 percent per decade).

The mechanisms causing changes in Antarctic sea ice conditions investigated during IPY remain under intense investigation. Oceanographic factors appear to be important, and theories explaining the observed differences between the significant sea ice retreats in the Antarctic Peninsula region and the slight advances in the seas off eastern Antarctic have been suggested. One particularly interesting theory includes couplings between processes involving stratospheric ozone, changes in atmospheric pressure patterns, oceanographic upwelling, and sea ice distributions (Sigmond et al., 2011; Thompson and Solomon, 2002). Most of these linkages have been supported by an analysis of melt features on the Pine Island Glacier ice shelf (Bindschadler et al., 2011). Interesting correlations have also recently been explored between Antarctic sea ice extent changes and variations in the state of the Southern Annual Mode and the El Nino–Southern Oscillation indexes of atmospheric pressure pattern variability (Stammerjohn et al., 2008; Figure 3.9).

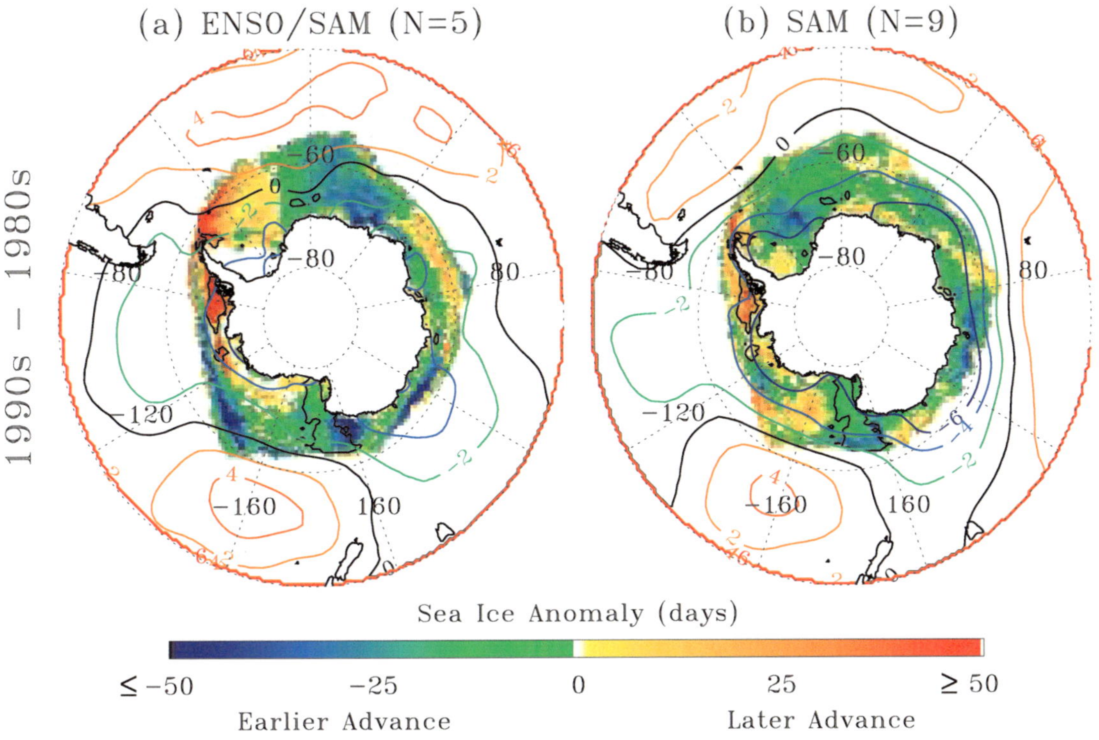


FIGURE 3.9 Decadal composite differences of sea ice advance and retreat with emphasis on ENSO and Southern Annular mode. SOURCE: Stammerjohn et al., 2008.

## MARINE ECOSYSTEMS IN A WARMING WORLD

While disappearing sea ice in the Arctic, portions of Antarctic, and glaciers are among the most visible surface indications that our planet is warming, IPY research also served to highlight the vulnerability of Arctic and Antarctic polar marine ecosystems to warming and to provide benchmarks of present physical and biological ocean conditions against which changes can be quantified. In contrast to earlier international years where polar biology and ecology were nearly ignored, research conducted during IPY 2007-2008 clearly demonstrates that such changes are having a serious impact at all trophic levels—from microorganisms to top predators (Grebmeier, 2012; NRC, 2011a,b). It remains a major forecasting challenge to understand the ecological, biogeochemical, and socioeconomic implications and broader impacts of these changes and predict their future courses as warming and sea ice loss proceed over the next few decades. New international observation systems put in place during IPY (e.g., Sustaining Arctic Observing Networks [SAON]; and the U.S. Sea Ice Mass Balance in the Antarctic, IPY/ASEP, and SO GLOBEC; also see Chapter 4) are providing data critical for determining the pace of future change and identifying the complex mechanisms driving ecosystem modifications in both the Arctic (western Arctic, e.g., Chukchi Sea) and the Antarctic (western Antarctic peninsula).

### Arctic Discoveries

As a consequence of a warming world and associated changes in sea ice and ocean currents, Arctic scientists anticipate significant shifts and reorganization of marine ecosystems. The dramatic sea ice loss of 2007 was depicted in images of polar bears and walruses stranded as their habitat literally melted beneath them, and there is evidence of reduced body size in polar bears caused by declining sea ice (e.g., Rode et al., 2010). Projections of future sea ice distributions (e.g., Overland, 2011) indicate that summer ice cover is likely to remain for longest in the region north of Canada and Greenland where the oldest and thickest ice now occurs (Figure 3.8). This area may become a refuge (Pfirman et al., 2009) for ice-associated species such as seals (e.g., Kelly et al., 2010) and the polar bears that depend on them (e.g., Durner et al., 2009).

Significant shifts and reorganization of marine ecosystems are now anticipated by Arctic scientists as a consequence of environmental warming and associated changes in ocean currents. In the Norwegian, Barents, and Chukchi Seas (Figure 3.10), water temperatures have reached the highest values ever recorded (Dickson and Fahrbach, 2011; IPCC, 2007a). The Atlantic water layer that penetrates at depth into the Arctic Basin has been warming and changing salinity, especially from the Laptev Sea into the Canada Basin (Dmitrenko et al., 2009, 2008; Polyakov et al., 2007) consistent with earlier observations first documented by the Surface Heat Budget of the Arctic Ocean Program (Morison et al., 2000). At the same time, in the Pacific sector of the Arctic, North Pacific and Bering Sea water masses penetrating into the Arctic via the Bering Strait are also warming, triggering significant sea ice melt while becoming an important factor leading to thinning of the seasonal ice pack across the western Arctic (Steele et al., 2008; Woodgate et al., 2010).

For the eastern Arctic to the east and west of Svalbard, warm Atlantic water enters the Arctic Basin from the Greenland and Barents Seas. Here recent studies support the idea that the sensible heat carried by this water causes the southward moving ice to melt, resulting in a decrease in the salinity and density of the upper part of the water column and thereby contributing to

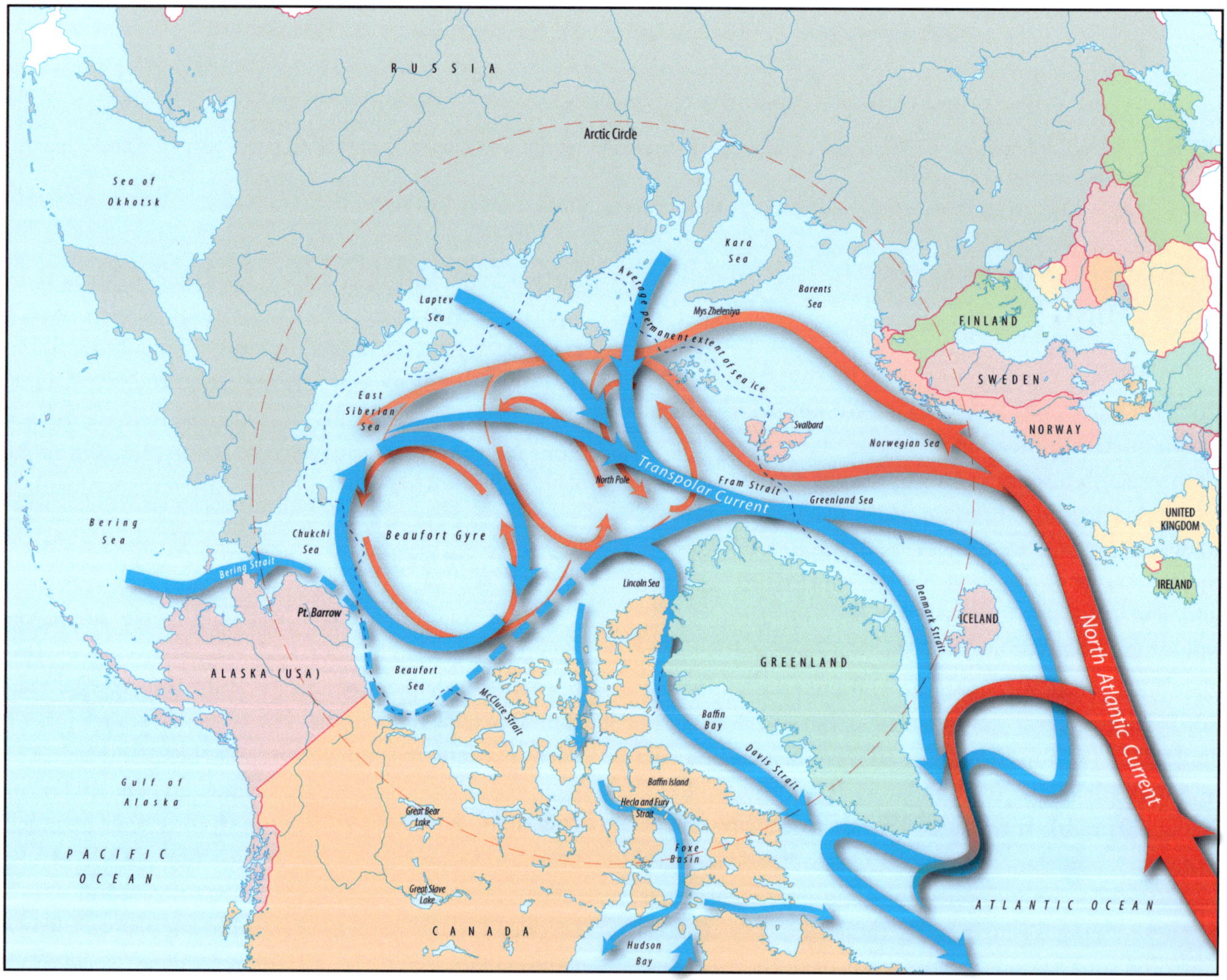


FIGURE 3.10 Cold, relatively fresh water from the Pacific Ocean enters the Arctic Ocean through the Bering Strait (blue arrows). It is swept into the Beaufort Gyre and exits into the North Atlantic Ocean through three gateways (Fram, Davis, and Hudson Straits). Red arrows indicate warmer, more saline water from the Atlantic Ocean. SOURCE: Jack Cook, Woods Hole Oceanographic Institution.

the development of a stable low salinity layer below the surviving sea ice. However, most of the Atlantic water is subducted beneath this low salinity layer to a depth where its heat content appears to have little effect on the growth and decay of the overlying ice (Maslowski et al., 2001; Polyakov et al., 2011).

The Bering and Chukchi Seas are key areas in biological terms because they are among the most productive waters on Earth; half of the marine harvest of the United States currently comes from the Bering Sea alone (King et al., 2005). Spatial and temporal shifts in the freeze-up and melt-back of seasonal ice in the Bering Strait region and western Arctic are causing measureable shifts in many marine species (Figure 3.11). Earlier ice breakup changes the timing and intensity of the ice edge bloom because of its association with seasonal changes in sunlight and day length. Such changes in primary productivity and nutrient utilization at the base of the food chain are causing reorganization of zooplankton communities, which can result in an increase in zooplankton abundance and biomass (Arrigo et al., 2008; Matsuno and Yamaguchi, 2010). Moreover, reductions in sea ice appear to be propelling the northward migration of larger warm-adapted Pacific species with faster growth rates (Grebmeier, 2012). Changes in surface water productivity drive changes throughout the water column, notably in benthic communities, which are prey for larger species including ducks, bearded seals, walrus, and whales (Bluhm et al., 2009; Grebmeier et al., 2006).

While the Beaufort and Chukchi Seas north of the Bering Strait largely support important fisheries for subsistence hunters in coastal communities, the ongoing transformation of these marine ecosystems suggests that commercial operations will eventually respond and push to follow their catch's migration. The northward migration of commercially important pelagic and ground fish will likely keep pace with the future retreat of the summer sea ice edge, effecting changes to marine management strategies (see Chapter 5). Some commercially fished species traditionally found in the southern Bering Sea, including walleye pollack, Pacific cod, and Bering flounder, are today observed in the Beaufort Sea along with commercially sized snow crabs (Grebmeier, 2012; Rand and Logerwell, 2011). The Distributed Biological Observatory initiated during IPY is being developed across the Pacific Arctic ocean gateway as a new approach to monitor significant range extensions and ecosystem transformation into the near future. This complements other physical and biological observatories in place as part of the new SAON[5] across the Arctic basin and marginal seas.

## Antarctic Discoveries

Ecosystem change is also apparent across the high latitudes of the Southern Hemisphere and was further documented during IPY. The Southern Ocean Global Ocean Ecosystems Dynamics (SO GLOBEC) program is an international multidisciplinary effort designed to provide understanding of the physical and biological factors that influence growth, reproduction, recruitment, and overwintering survival of Antarctic krill (*Euphausia superba*), a key species in the Antarctic food web (Hofmann et al., 2011; Figure 3.12). The end-to-end food web approach that included predators and competitors of Antarctic krill, as well as the influence of habitat, makes this program unique and allowed comparative studies across ecosystems in the western Antarctic Peninsula (WAP) region (U.S. program), East Antarctica (Australian program), the Lazarev Sea (German program), and South Georgia (UK program).

The SO GLOBEC program continued through IPY to provide new insights into the functioning of the Antarctic marine food web, especially in the winter. Primary discoveries from this program showed that (1) biological distributions in many regions of the Southern Ocean, particularly the western Antarctic Peninsula region, are structured by hydrography, circulation, and sea ice; (2) Antarctic krill use a range of overwintering strategies; (3) biological hot spots (regions with enhanced predator abundance relative to other areas) exist; (4) distributions of top predators (e.g. fish, penguins, seals, and cetaceans) are correlated with habitat structure and prey availability; and (5) sympatric krill predators (crabeater seals, minke whales, humpback whales) appear to have little niche overlap (Friedlaender et al., 2011). Many SO GLOBEC studies provided the first winter observations of top predator condition, distribution, and habitat use.

The synthesis and integration of the SO GLOBEC data sets focused on comparative studies within and

[5] *www.arcticobserving.org/.*

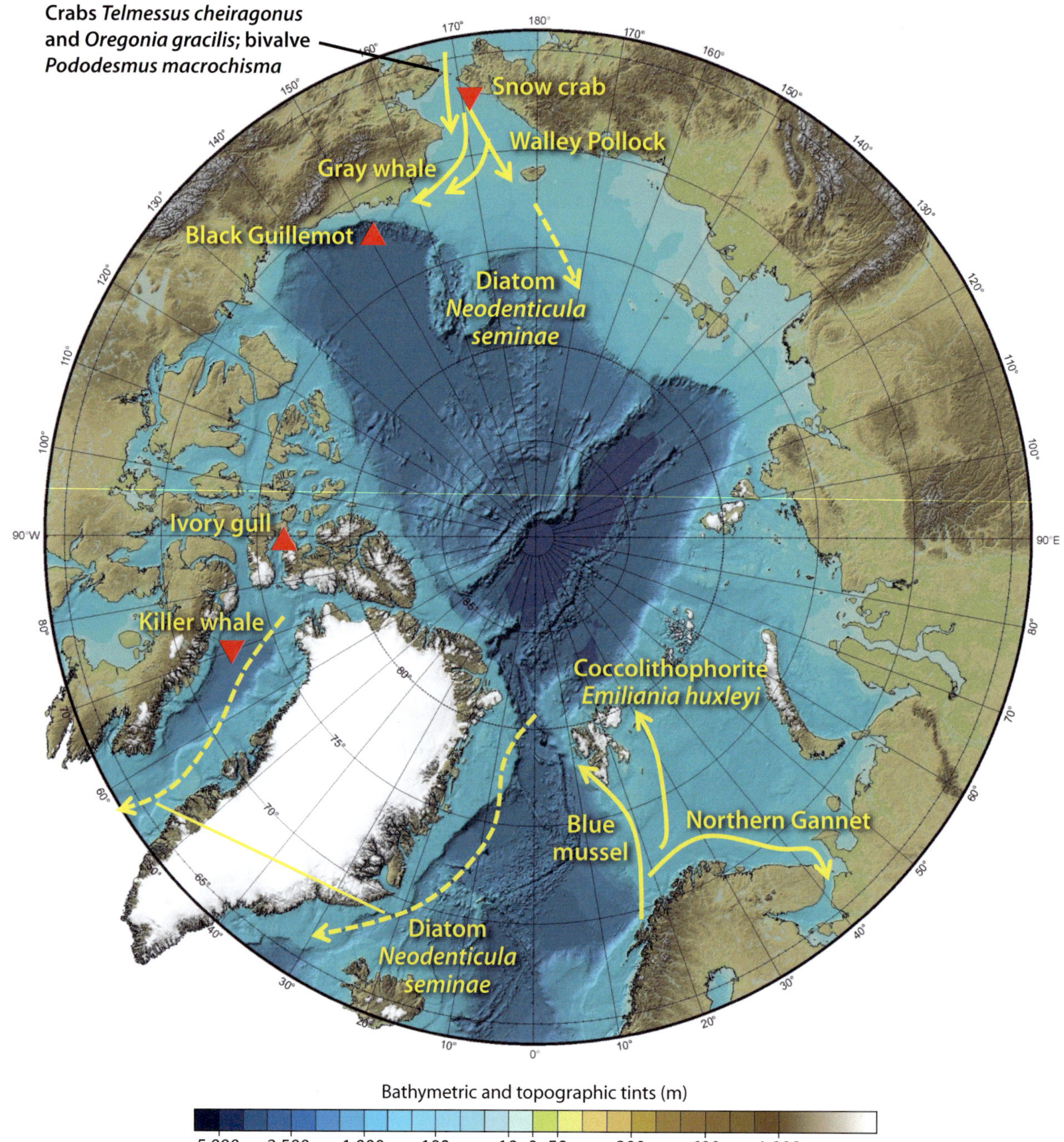


FIGURE 3.11 Map of examples of change in species distributions or population size or sightings that have been attributed to global climate change. Yellow arrows show the general direction of the species range change, but are not meant to suggest exact pathways. Red triangles indicate increases or decreases in population numbers or sightings. SOURCE: Bluhm et al., 2011.

between regions and included a significant modeling component. Circulation modeling studies of Circumpolar Deep Water intrusions onto seas west of the Antarctic Peninsula and in the Ross Sea showed differences in vertical mixing of this water mass that have implications for different basal melting rates of ice shelves and nutrient supply in the two regions (Dinniman et al., 2011). Numerical Lagrangian particle tracking studies showed the importance of the circulation in determining shelf residence times of water and biota, transport pathways, and retention regions, which have implications for connectivity and recruitment of Antarctic krill populations (Piñones et al., 2011; Wiebe et al., 2011). An important contribution from SO GLOBEC

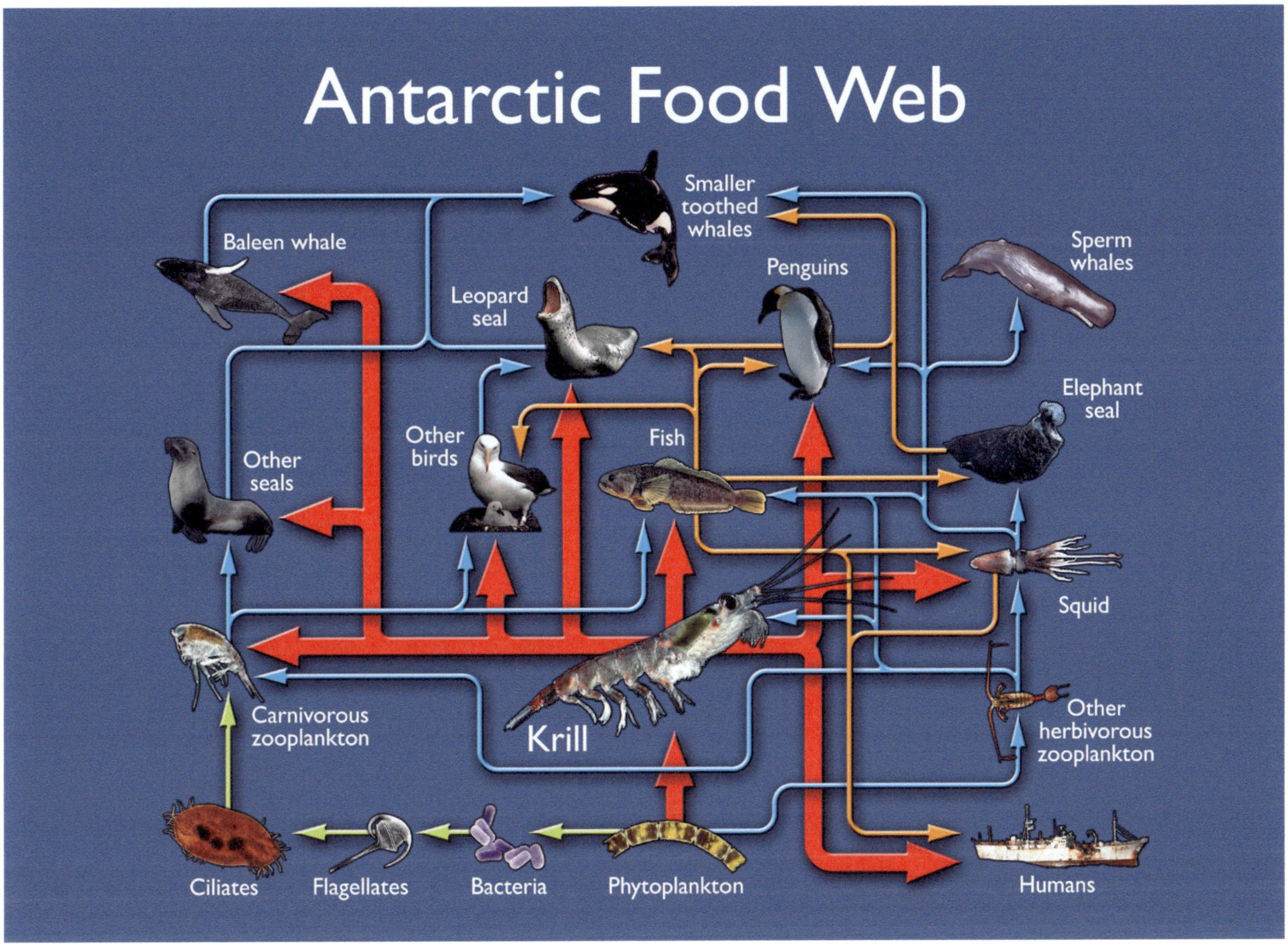


FIGURE 3.12 Antarctic food web showing krill at the center. SOURCE: British Antarctic Survey.

during IPY was demonstrating that data acquired from conductivity-temperature-depth satellite-data relay loggers deployed on crabeater (*Lobodon carcinophagus*) and southern elephant (*Mirounga leonine*) seals (Figure 3.13) could be used to study upper ocean properties such as seasonal variability in heat content (Costa et al., 2008), physical profiling and ocean state estimation (Mazloff and Wunsch, 2010), and sea ice production (Meredith et al., 2011). Tagged animals were an integral part of several SO GLOBEC studies.

The Amundsen Sea Embayment Project (ASEP[6]) has continued since the IPY period as a multidisciplinary effort designed to study the upwelling of relatively warm Circumpolar Deep Water onto the Amundsen Sea continental shelf, how this relates to atmospheric forcing and bottom bathymetry, and how the warm waters interact with both glacial and sea ice. This region of the West Antarctic ice sheet is losing ice faster, perhaps by orders of magnitude, than other Antarctic ice sheet regions, excluding those on the Antarctic peninsula (Pritchard et al., 2009). Thus, continued monitoring of the West Antarctic ice sheet is of crucial importance. As in other parts of the Antarctic, submarine depressions that cross the continental shelf provide conduits for the across-shelf transport of Circumpolar Deep Water to the cavity under the floating extension of the ice shelf (Jacobs et al., 2011). The hydrographic measurements made as part of ASEP during IPY are now being used to provide quantitative estimates of the rate of glacial ice melt by the Circumpolar Deep Water that intrudes across the Amundsen Sea continental shelf. These rates are critical to providing estimates of the potential for sea level rise caused by melting of the West Antarctic ice sheet (Rignot et al., 2011).

[6] *http://www.antarctica.ac.uk/staff-profiles/webspace/dutrieux/Research/ASEdatabase/.*

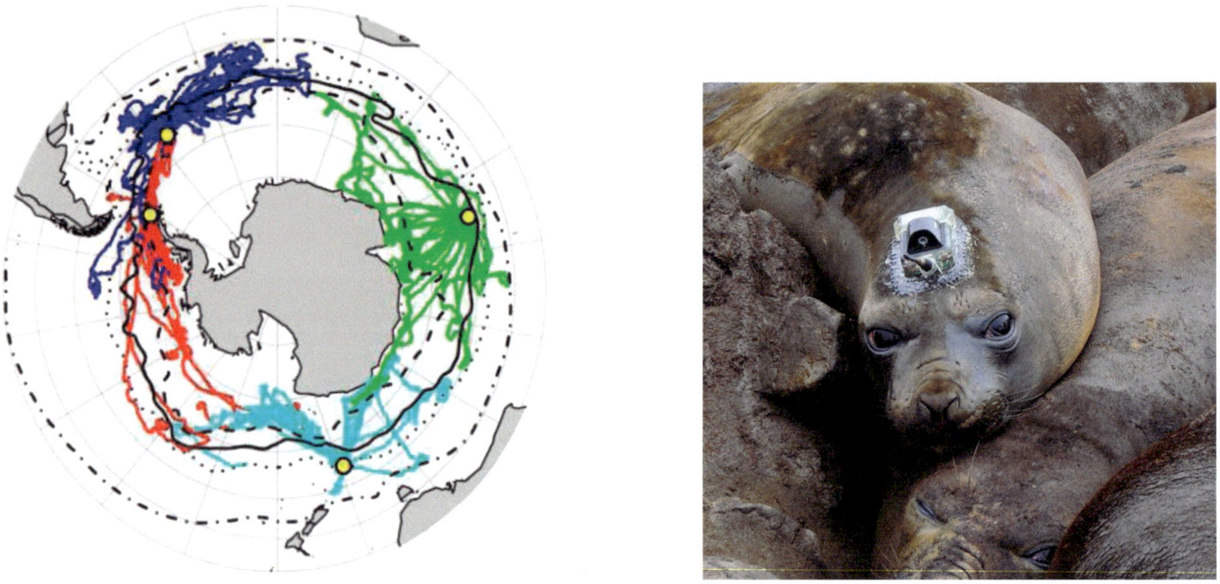

FIGURE 3.13 Southern elephant seal tracks from animal-mounted CTD-SRDL sensors (left) and an elephant seal with a transmitter to collect data on conductivity, temperature, and depth, covering a large geographic area (including areas that would be otherwise difficult to reach) (right). SOURCES: (left) Biuw et al., 2007; (right) D. Costa.

## MARINE CARBON CYCLING AND OCEAN ACIDIFICATION

The Arctic marine carbon cycle and exchange of carbon dioxide ($CO_2$) between the ocean and atmosphere appear particularly sensitive to environmental changes, including sea ice loss, warming, changes in seasonal marine phytoplankton primary production, changes in ocean circulation, and freshwater inputs (Bates et al., 2011). Carbon cycling is especially vigorous in the spring when northward retreat of the ice edge results in high productivity within the freshly exposed surface waters (Gradinger, 2009; Lee et al., 2011).

Several IPY studies have shown that the Arctic Ocean is currently a significant sink for atmospheric $CO_2$ (Bates and Mathis, 2009; Bates et al., 2011). In the western Arctic Ocean, high rates of primary production in the Chukchi Sea (Figure 3.14) make the region a strong seasonal ocean sink for $CO_2$ that is partially compensated by outgassing of $CO_2$ from the East Siberian Sea shelf region, where large fluxes of easily altered, terrigenous organic carbon is remineralized in the coastal zone (Mathis et al., 2009). In the near term, further sea ice loss and increases in phytoplankton growth rates are expected to cause a limited net increase in the uptake of $CO_2$ by Arctic surface waters. Recent studies suggest that this enhanced uptake will be short-lived with surface waters rapidly warming and equilibrating with the atmosphere. Furthermore, release of large stores of carbon from the surrounding Arctic landmasses through rivers into the Arctic Ocean and further warming over the next century may alter the Arctic from a $CO_2$ sink to source over the next century (Holmes et al., 2011). This ambiguity in even the sign of the Arctic's future role in the carbon cycle underscores the value of IPY research in documenting the present state and providing data to help unravel both the magnitude and the sensitivity of components of the carbon cycle to environmental changes.

Fossil fuel combustion and industrial processes release over 6 billion metric tons of carbon into the atmosphere each year, and $CO_2$ concentrations in the atmosphere are projected to continue to rise (IPCC, 2007a). The consequences of these greenhouse gas emissions are often discussed in terms of rising global temperatures, but global air temperature rise is not the only threat from increased atmospheric concentrations of $CO_2$. The decrease in seawater pH due to the uptake of anthropogenic $CO_2$ has been termed ocean acidification. Ocean acidification, which occurs when $CO_2$ in the atmosphere reacts with water to create carbonic acid, has increased ocean acidity by 30 percent

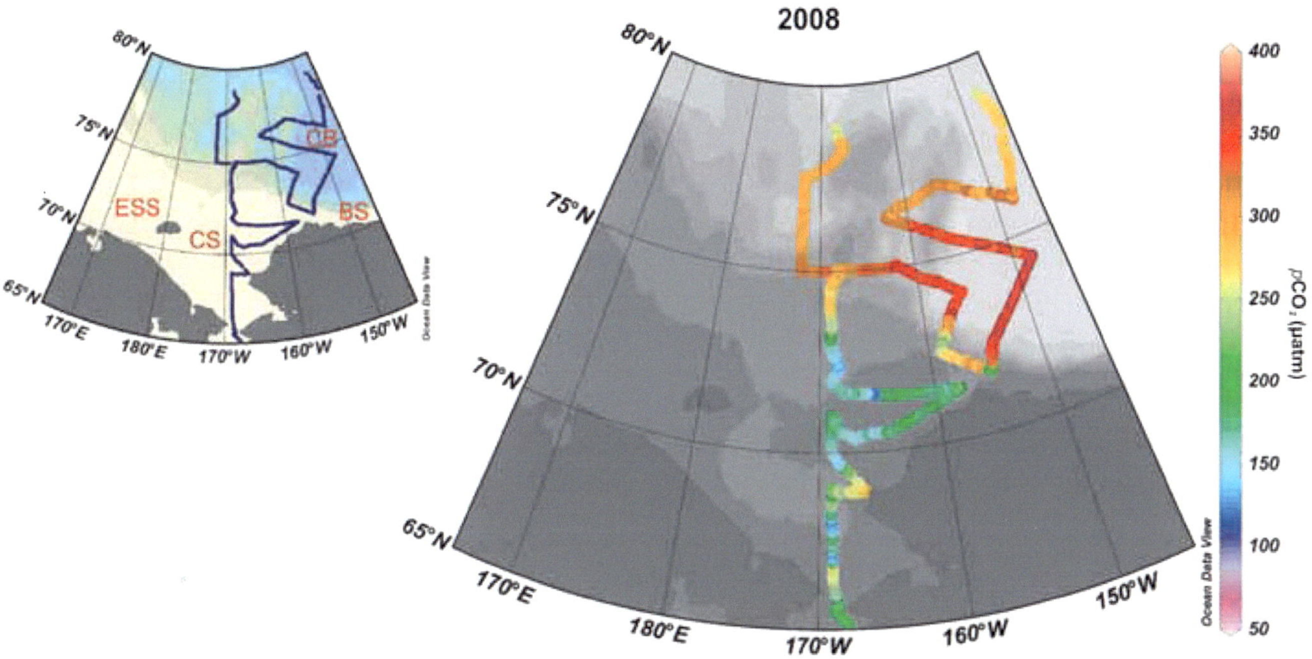


FIGURE 3.14 Surface distributions of seawater partial pressure of $CO_2$ ($pCO_2$ in µatm) during the summer 2008 China National Arctic Research Expedition (CHINARE). Data were collected using a National Oceanic and Atmospheric Administration Atlantic Oceanographic and Meteorological Laboratory (AOML) underway $CO_2$ system deployed on the icebreaker *Xuelong*. SOURCE: Bates et al., 2011.

(Doney, 2006). Compared to the mean annual global ocean $CO_2$ uptake of approximately 1400 Tg C per year (Takahashi et al., 2009), the Arctic Ocean $CO_2$ sink potentially contributes 5-14 percent to the global balance of $CO_2$ sinks and sources (Bates and Mathis, 2009); the Southern Ocean uptake of anthropogenic $CO_2$ is large, but storage there is relatively small (Caldeira and Duffy, 2000). Although the chemistry of this effect is well understood and not much debated, the full consequences of ocean acidification for marine ecosystems and human well-being are only beginning to be revealed (NRC, 2010b).

IPY-related research on ocean acidification has brought to light the possibility that polar waters are likely to experience the effects of ocean acidification before most other regions of the world. The uptake of anthropogenic $CO_2$ has already decreased surface water pH by 0.1 units and under current scenarios of possible future human emissions (IPCC, 2007a) a further decrease in seawater pH by 0.3 to 0.5 units over the next century is possible, with the Arctic Ocean likely experiencing these changes on shorter time scales. Various studies have shown that the Southern Ocean may become acidic enough to begin to experience aragonite undersaturation, which would weaken the ability of calcifying organisms to form calcium carbonate shells, within this century (Cao and Caldeira, 2008; McNeil and Matear, 2008).

## POLAR ATMOSPHERIC OBSERVATIONS AND LOWER LATITUDE IMPACTS

The intimate connection of the polar regions with the rest of the globe is most evident in the atmosphere, where air from lower latitudes can be transported to the poles in a matter of days. Numerous projects in IPY explored these connections in a variety of ways—in the lower and upper atmosphere; in chemistry, physics, and dynamics; and in the Arctic and the Antarctic regions. IPY research on the atmosphere highlighted how the poles are vulnerable to what happens at low latitudes. For example, agricultural activities and forest fires in Asia and North America largely drive the annual formation of Arctic haze. In addition, pollution transported to the Arctic, such as persistent organic pollutants (POPs), will be deposited there. Similarly, IPY research also highlighted how the rest of the globe is affected by what happens at the poles: for example, as noted above, Arctic circulations can influence weather patterns across the Northern Hemisphere, and the competing effects of climate change and ozone depletion in the Antarctic stratosphere have altered precipitation and temperature patterns in the Southern Hemisphere. Our understanding of these connections improved during IPY, but there is more to learn, and continued monitoring is necessary to understand how these connections will evolve as climate changes.

Two major objectives of IPY were the establishment of observing networks and increased sharing of data among a number of countries. Projects focused on observations of the polar atmosphere, such as IASOA (International Arctic Systems for Observing the Atmosphere), SPARC-IPY (Stratospheric Processes And their Role in Climate[7]), SCSCS (Spitsbergen Climate System Current Status), and COMPASS (Comprehensive Meteorological data set of active IPY Antarctic measurement phase for Scientific and applied Studies[8]) showed that observing common parameters in multiple locations allows for better understanding of processes that affect entire polar region, whether in the Arctic, the Antarctic, or both.

## Arctic Highlights

Black carbon has a complex role in the atmosphere. As an air pollutant it has clear negative impacts on human health, but as a climate forcing agent, it has competing effects directly through absorption of radiation, indirectly through cloud formation, and through changes to the albedo of snow and ice when deposited on the surface. The IPY provided an opportunity to characterize sources, sinks, and transport pathways in the Arctic for black carbon and other short-lived species in a way that has never before been achieved. The coordinated international efforts under POLARCAT[9] created the ability to take a "snapshot" of synoptic measurements in both spring and summer in a truly pan-Arctic capacity. From these observations, scientists were able to (a) identify key source regions for Arctic black carbon; (b) assess the potential for high-latitude ozone chemistry to influence the troposphere at lower latitudes; (c) gain a better understanding of the dynamics of sea ice and the influence on tropospheric concentrations of bromine, ozone, and mercury; and (d) establish the highest possible quality measurements from which to benchmark model simulations, satellite retrievals, and future Arctic characterizations. None of this would have been feasible without the internationally coordinated research investment that IPY provided.

Arctic haze has been known to have much of its origins from Northern Hemisphere anthropogenic pollution, biomass burning, and dust. IPY research has more clearly shown the influence of Northern Hemisphere pollution transport (particularly northeastern Europe and Siberia) on the seasonal background, as well as the episodic transport of specific pollution layers, from a wide variety of Northern Hemisphere locations including events associated with agricultural and forest fires (Brock et al., 2011; Figure 3.15). Thus, the Arctic haze has both chronic and episodic characteristics. This Arctic aerosol loading plays a direct role

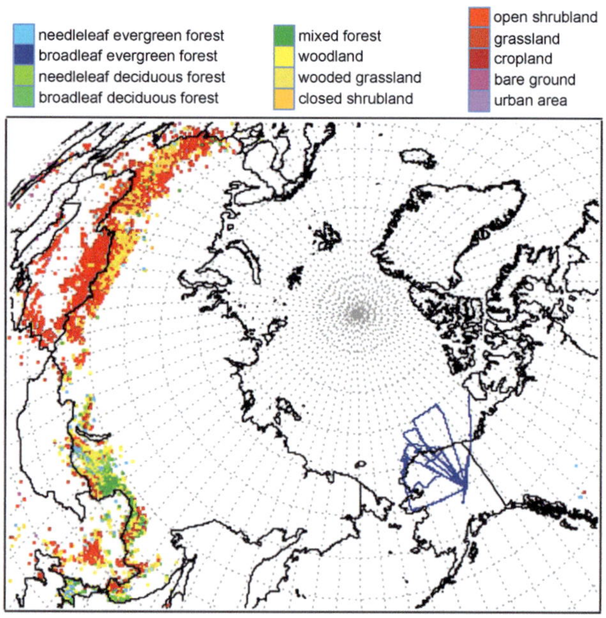


FIGURE 3.15 Arctic haze has been known to have much of its origins from Northern Hemisphere anthropogenic pollution, biomass burning, and dust. During IPY, fires were shown to contribute to Arctic haze more significantly than previously estimated. This map shows fire locations for April 2008 determined from MODIS (NASA/University of Maryland, 2002), color-coded by the underlying vegetation type (Hansen et al., 2000). SOURCE: Brock et al., 2011.

---

[7] *http://www.atmosp.physics.utoronto.ca/SPARC-IPY/.*

[8] *http://classic.ipy.org/development/eoi/proposal-details.php?id=267.*

[9] *http://www.polarcat.no/*; the United States played a significant role in this through the ARCTAS (*http://www.espo.nasa.gov/arctas/*), ISDAC (*http://acrf-campaign.arm.gov/isdac/*), ARCPAC (*http://www.esrl.noaa.gov/csd/arcpac/*), and ICEALOT (*http://saga.pmel.noaa.gov/Field/icealot/*) field projects.

in the Arctic climate system by producing heating aloft in the atmosphere and cooling at the surface.

A significant concern with this increase is that the increased loading of light-absorbing aerosol (black carbon and dust) to the surface could lead to increased melting of snow and ice, exacerbating the effects of warming. However, IPY research shows that the black carbon content of snow in the Arctic has not been increasing in recent decades and therefore cannot be a major contributor to the recent Arctic sea ice loss (Doherty et al., 2010).

New insights were also gained into the behavior of ozone in the Arctic. Ozone depletion events (ODEs) have been observed in the past, where elevated bromine levels in the Arctic boundary layer can drive these ODEs. Elevated bromine monoxide (BrO) "hotspots" were observed by two satellites (OMI and GOME-2) during IPY. Previously it had been thought that the source of this bromine was from the surface, but during IPY these elevated bromine levels were found to be correlated with tropopause depressions, indicating that the source of bromine is from the stratosphere (Jacob et al., 2010; Salawitch et al., 2011). This is a new understanding of ozone loss in the Arctic and generally on the behavior of the Arctic atmosphere.

POLARCAT research at Summit, Greenland, confirmed that active bromine chemistry is occurring in and just above the snowpack despite the great vertical and horizontal separation from strong sources of bromide. Activation of the trace amounts of bromide in the snow on top of the Greenland ice sheet is able to support vigorous bromine chemistry that was shown to perturb the cycling of hydroxyl, nitrogen oxides, and mercury. This research also included development of the first one-dimensional coupled air/snow chemistry model able to simulate observed variations of BrO and NO above the snow based on production and release within the surface layers of the snowpack (see Brooks et al., 2011; Dibb et al., 2010; Liao et al., 2011; Stutz et al., 2011; Thomas et al., 2011).

The Northern Hemisphere winters of IPY corresponded to years where there were major disturbances of the Arctic polar vortex (mentioned above). In both winters, large planetary waves broke in the stratosphere and disrupted the circulation. These waves caused major displacement of the vortex in 2007-2008 and caused a splitting of the vortex in 2008-2009. These disruptions are of interest for two distinct reasons: (1) the disruption of the vortex mixes cold vortex air with warmer air from lower latitudes, which impedes polar stratospheric cloud formation and thus causes less ozone depletion; and (2) the larger disruption of the circulation acts as a natural laboratory for understanding how large-scale and small-scale waves impact the general circulation. IPY research involving networks of lidar observations characterized these behaviors across the Arctic (Farahani et al., 2009; Thurairajah et al., 2010), where small-scale waves were more severely impacted during the winters with the larger disturbance (2008-2009). The observations provide a benchmark for model studies of how wave-driven variability impacts the general circulation.

### Antarctic Highlights

The Antarctic ozone hole and global air temperature rise both affect Southern Hemisphere circulation. Climate model integrations of the 20th century including ozone-only forcing show diminished sea ice and a warming of the Southern Ocean surface air temperatures (Sigmond et al., 2011). It also appears clear from these modeling studies as well as the work of Arblaster et al. (2011) that the important issue here is the dynamic response to the ozone-induced cooling and the associated surface climate response to the circulation change. Current studies do not provide a ready explanation for the observed expansion of sea ice extent in the eastern portions of the Southern Ocean (Thompson et al., 2011). Over the past several decades, the ozone hole has shifted the extratropical westerly jet poleward, which has resulted in increased precipitation in the Southern Hemisphere subtropics (Kang et al., 2011; Figure 3.16). The relative contribution of ozone depletion and greenhouse gas warming will change as the ozone hole recovers and climate change intensifies, so continual monitoring will be needed to understand how these separate influences evolve and their net effect on Southern Hemisphere circulation in the coming years.

Closer to the surface, as mentioned above, warming has been observed over the Antarctic Peninsula, but observations of temperature are still relatively sparse over much of the Antarctic continent. Recent analyses during the IPY time frame suggest that this warming

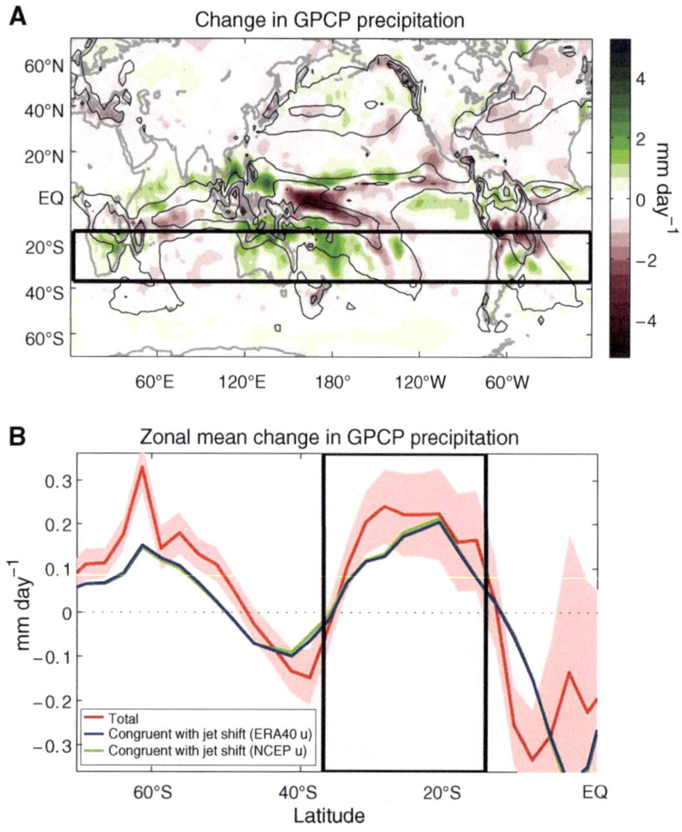


FIGURE 3.16 The polar regions have an important role to play in the subtropical hydrological cycle. A poleward shift in the extratropical westerly jet caused by the Antarctic ozone hole has led to precipitation changes in the Southern Hemisphere subtropics. (A) Precipitation change based on GPCP data, calculated from the linear trend multiplied by 22 years. Black contours show the mean precipitation for 1979-1983, with contour interval of 3 mm day-1. The area within the black box exhibits a moistening trend during the period of ozone depletion between 1979 and 2000 in austral summer (December to February mean). (B) This result can also be seen in the zonal-mean. The precipitation change is indicated by the red line with 95 percent confidence interval (red shading), and the change congruent with a change in the latitude of the westerly jet obtained from ERA40 (blue line) and NCEP/NCAR (green line) reanalysis data. SOURCE: Kang et al., 2011.

may be more widespread than just the peninsula (Steig et al., 2009).

## GEOSPACE AND SPACE WEATHER HIGHLIGHTS

The decades since the advent of spaceflight have witnessed the increasing importance and relevance of the Earth's space environment—geospace (or near-Earth space including the upper regions of the atmosphere, magnetosphere, and ionosphere). This region plays host to the myriad processes that are largely driven by solar forcing and that involve complex interactions with the Earth's magnetic field, which are collectively referred to as space weather. For example, in the polar regions the solar wind energy is transferred into disturbances in the Earth's magnetosphere and ionosphere that can affect and even disrupt technologies in space and on the surface, affecting communications for polar-route aircraft and exposing passengers and crew to high levels of particulate radiation at airline altitudes.

The science of space plasma physics has matured to the level of being able to both describe and predict many of the interactions in geospace. An overarching goal of IPY was to fold current understanding into a more comprehensive computational framework that could be used to analyze and predict the properties of this system. To realize such a goal, coordinated ground- and space-based measurements were necessary both to resolve outstanding problems and to provide key data assimilation parameters. During the IPY time frame, many countries, including the United States, invested in autonomous observatory ground-based infrastructure that substantially increased the number of geospace observations in Antarctica (Mende et al., 2009; Musko et al., 2009; Figure 3.17). IPY thus allowed more comprehensive synoptic measurements of the geospace environment. The modulation of the magnetosphere's open-closed boundary (OCB) during periods of high-speed solar streams was inferred from fluxgate magnetometer instrumentation covering a wide range of high-latitude geomagnetic latitudes and longitudes across Antarctica (Urban et al., 2011). Researchers also demonstrated how the OCB can be deduced in real time from synoptic data sets, predicting

FIGURE 3.17 During IPY, geospace observation stations, like the one pictured here in Antarctica, substantially increased the amount of data collected. SOURCE: A. Stillinger.

intense radiation patches and high-frequency radio communications disruptions.

During IPY, scientists also conducted studies revealing new information about the linkage between solar activity and temperatures in polar regions. The study by Seppälä et al. (2009) revealed that surface temperatures in certain high-latitude regions varied by ~4-5 degrees during enhanced geomagnetic activity as compared to temperatures measured during quiet geomagnetic conditions. Although the authors could not conclusively show that the polar surface air temperature patterns are physically linked by geomagnetic activity, they did demonstrate that geomagnetic activity likely plays a role in modulating wintertime surface air temperatures.

Images of the aurora taken simultaneously in the Northern and the Southern Hemispheres during IPY revealed indisputable evidence that the auroras in the two hemispheres can be totally asymmetric (Laundal and Østgaard, 2009; Figure 3.18). Before this study was published, it was commonly assumed that the aurora borealis (Northern Hemisphere) and aurora australis (Southern Hemisphere) were approximately mirror images of each other because the charged particles producing the aurora follow magnetic field lines connecting the two hemispheres, and the particles were believed to be evenly distributed between the two hemispheres, from the major source region in the equatorial plane of the magnetosphere. The asymmetry is interpreted in terms of interhemispheric electrical currents related to differing seasons, which had been predicted but hitherto had not been seen.

## TERRESTRIAL EARTH SYSTEMS AND PERMAFROST

Current warming of the planet is contributing to changes in terrestrial environments that have large impacts on society. Vivid images leading into the IPY years depicted villages in Alaska falling into the sea due to coastal erosion, broken pipelines, and crumbling runways resulting from the thaw of permafrost (ACIA, 2005). These images at the time conveyed that climate change may be only impacting Arctic infrastructure and Arctic people. Yet during the IPY years the world came

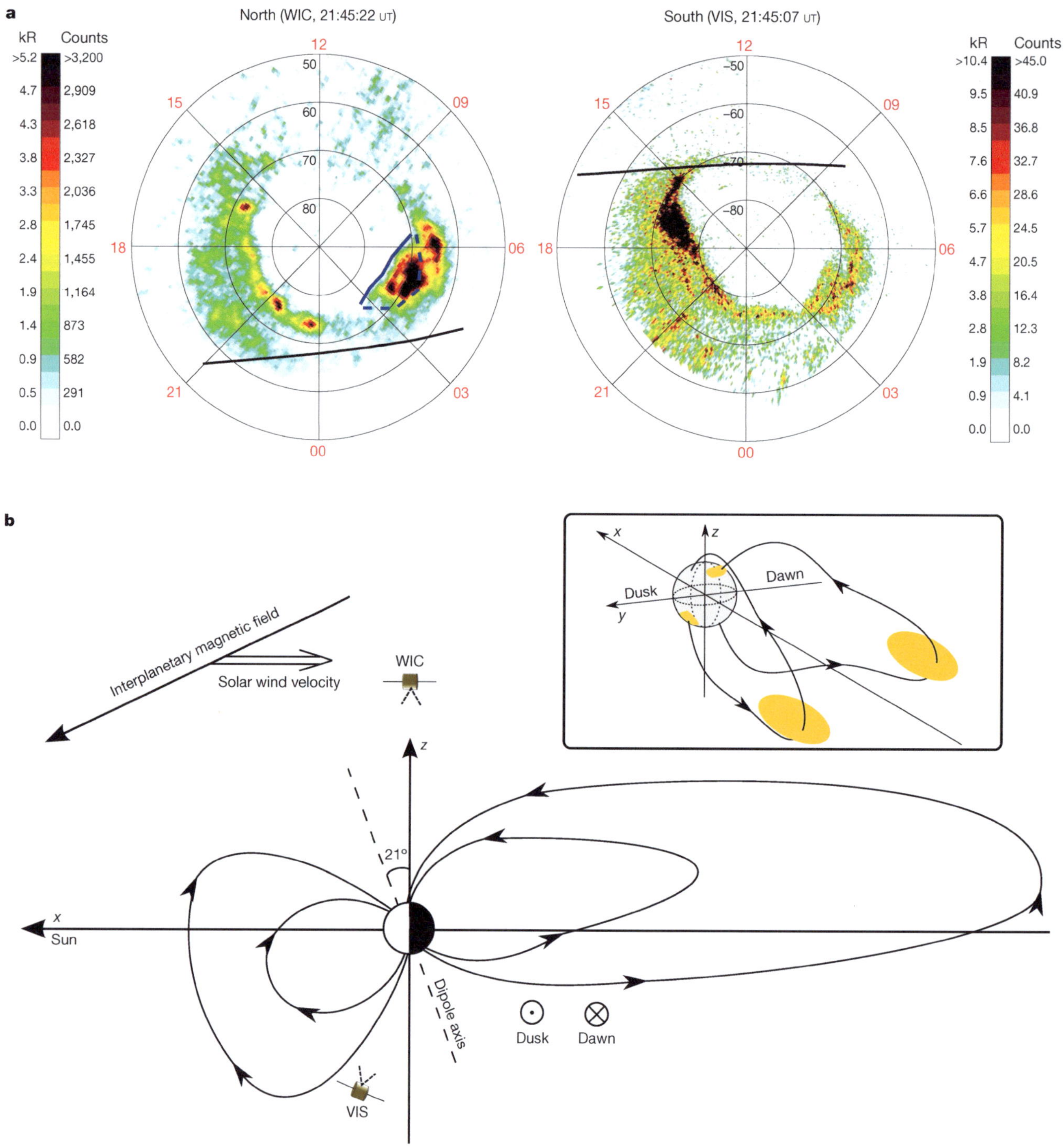


FIGURE 3.18 Simultaneous ultraviolet auroras were observed in both hemispheres during IPY. The images from the Northern and Southern hemispheres reveal that the auroras in these two areas can be totally asymmetric. In fact, Laundal and Østgaard (2009) show that simultaneous images of the aurora on May 12, 2001, have completely different intensity distributions (a). The color bars show intensity in both counts and kilo-Rayleigh (kR). A conceptual presentation shows the seasonal conditions and geometry of the magnetosphere, and the orientation of the interplanetary magnetic field, measured by ACE (b). The interplanetary magnetic field moves with the solar wind velocity. The asymmetry is interpreted in terms of interhemispheric electrical currents related to differing seasons, a result that had only been predicted (not observed) prior to this study. The inset illustrates that the spots at northern dawn and southern dusk originate from completely different regions in the magnetosphere. SOURCE: Laundal and Østgaard, 2009.

to understand that warming impacts on permafrost also contribute to climate change globally, through feedbacks that affect the complex linkages and feedbacks between atmospheric, oceanic, and terrestrial systems.

Permafrost is soil and other subsurface materials that remain frozen year after year. It is widespread in the high-latitude regions (Figure 3.19). During IPY, nearly 350 new borehole observatories were drilled and instrumented in permafrost globally. This expanded the global network of permafrost observatories coordinated through the International Permafrost Association (Brown and Romanovsky, 2008) for continued observation and monitoring that will yield ongoing evidence for terrestrial-atmospheric climate feedbacks. These coordinated measurements demonstrate that permafrost is warming or thawing nearly everywhere throughout the circumpolar north (Figure 3.20).

Warming near-surface permafrost temperatures were observed in IPY, including observations reported for Svalbard (Isaksen et al., 2007) in the Atlantic sector of the Arctic Ocean north of Europe. The mean warming of 1-2°C over the last six to eight decades has been accelerating since 1999 at the rate of 0.6-0.7°C/decade. Air temperatures running 4-9°C above the 1961-1990 average were observed during the winter of 2005-2006 prior to IPY.

During IPY, the Carbon Pools in Permafrost Regions initiative gave rise to increased research on and increased public awareness of carbon reservoirs that have remained locked in permafrost during repeated cycles of glacial and interglacial change over the past few million years. It is now realized that the total below ground carbon pool in permafrost regions (~1,672 Pg C) is more than double the atmospheric carbon pool (~750

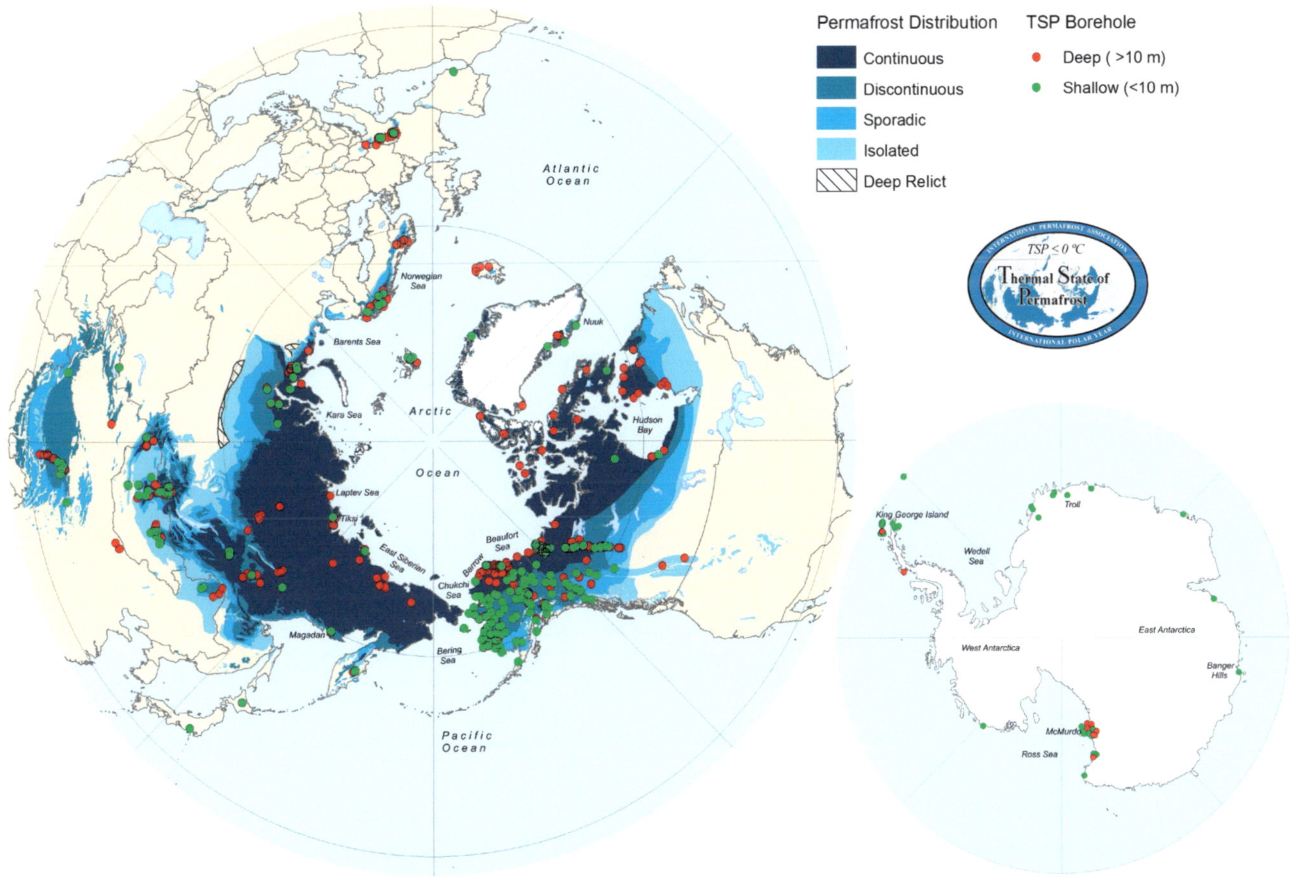


FIGURE 3.19 A large portion of the Arctic is underlain by permafrost (soil and other subsurface materials that remain frozen year after year). During IPY, nearly 350 new borehole observatories were drilled and instrumented in permafrost globally. On this map, the red and green symbols indicate borehole locations, darker shades indicate larger percentages of permanently frozen ground, and lighter shades (as well as the terms isolated and sporadic) refer to lower percentages of frozen ground. SOURCE: International Permafrost Association Standing Committee on Data Information and Communication, 2010.

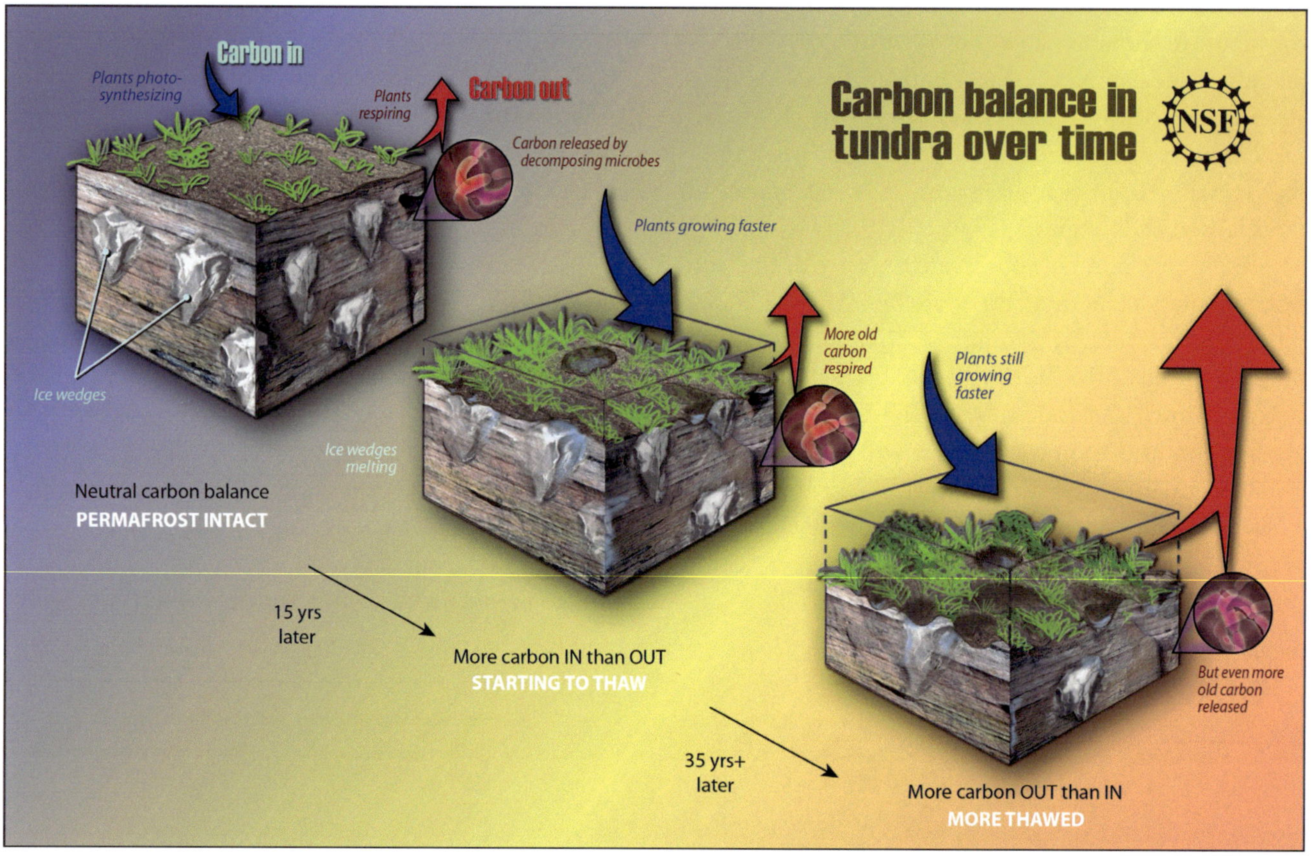


FIGURE 3.20 Scientists have demonstrated that, as permafrost begins to thaw, the amount of carbon removed from the atmosphere increases due to plant growth. As the permafrost continues to melt over time, however, the amount of carbon that is released increases and eventually overtakes the amount of carbon that plants remove. This process can contribute significant amounts of carbon to the atmosphere. SOURCE: Zina Deretsky, National Science Foundation.

Pg C), and three times larger than the total global forest biomass (~450 Pg C) (Tarnocai et al., 2009). Evidence is now emerging that deep-soil organic matter otherwise locked in permafrost in Alaska and northern Sweden is starting to be released (Dorrepaal et al., 2009; Schuur et al., 2009). Recent decades of greenhouse gas warming are causing the release of terrestrial-based carbon from permafrost areas at a rate that is abrupt compared to the glacial time scales involved in the creation of these terrestrial carbon reservoirs. Throughout human history, the land has been a weak source of carbon to the atmosphere while the ocean has been a sink accounting for 5-15 percent of the global carbon dioxide ($CO_2$) uptake from the atmosphere. Now, the potentially large positive climate feedback of additional carbon release to the atmosphere from thawing permafrost will change this longstanding balance and may overwhelm the ability of the ocean and other biomes to draw down atmospheric carbon, thereby contributing to even greater warming.

The Thermal State of Permafrost program yielded results that exemplify large differences in the existence and state of permafrost between sites that vary from continental to marine and from sea level to mountains (Christiansen et al., 2010; Romanovsky et al., 2010; Smith et al., 2010). The spatial variability and complexity of processes in permafrost point to the need for sustained observations and monitoring (Grosse et al., 2011; Jorgenson et al., 2010). Changes in permafrost have resulted in changes in hydrology, including expansion of some lakes (Parsekian et al., 2011) and shrinking of others (Roach et al., 2011). Lakes on the Seward Peninsula, Alaska, appear to be expanding as permafrost near boundaries degrades and subsides. Some lakes in the Yukon Watershed, Alaska, are shrinking as underlying permafrost thaws, allowing perched water to infiltrate to sublake groundwater.

Permafrost thaw is starting to have dramatic impacts on polar hydrologic systems, ecosystems, and the climate system. The Arctic terrestrial freshwater system is projected to transition from a surface water-dominated system to a groundwater-dominated system in the coming decades (Frey and McClelland, 2009). Consequent impacts from the changing hydrology and terrestrial ecosystems are changes to the aquatic ecosystem and biogeochemistry, including shifts in riverine concentrations of dissolved organic carbon (DOC), inorganic nutrients, cations, phosphate, and silicate. New satellite-based algorithms to map DOC and colored dissolved organic matter in the Kolyma River have improved understanding of the complexity of carbon delivery to the Arctic Ocean with strong mixing zones and downstream loading of carbon. These satellite observations may suggest that our current point-based estimates of carbon delivery to the Arctic Ocean may in fact be conservative (Griffin et al., 2011). This will have important effects on primary production and carbon cycling on polar shelves and oceans discussed earlier.

A synthesis of analyses on climate and hydrological processes in the Arctic reveal evidence that the Arctic hydrologic cycle is experiencing intensification (Rawlins et al., 2010). River discharge, precipitation, and evapotranspiration from both field measurements and GCM analyses exhibit positive trends, although significant positive trends above the 90 percent confidence level were not present for all data sets. Confidence in the trend of ocean flux rates is less certain because there are fewer long-term observations; however, ocean salinity and volume flux data suggest a decrease in freshwater outflow in recent decades. A decline in freshwater storage across the central Arctic Ocean and recognition that circulation through the Arctic Basin controls freshwater dynamics raises questions as to whether Arctic Ocean freshwater flows are intensifying. Although oceanic fluxes of freshwater are highly variable and consistent trends are difficult to verify, other processes in the Arctic freshwater cycle display positive trends over recent decades. These increases provide evidence that the Arctic hydrologic cycle is accelerating and inform us of the systemic links in terrestrial, marine, and atmospheric systems.

Warming of the planet creates regionally specific environmental changes. In years leading up to IPY, large changes in sea ice concentrations and thawing of permafrost in some Arctic coastal communities had been observed, but less attention had been focused on land-based ecosystems. Using data gathered by the Greening of the Arctic: Circumpolar Biomass IPY program, however, Bhatt et al. (2010) showed that land warming, sea ice decline, and greening of the Arctic are systemically linked. Summer sea ice concentrations have declined and summer land temperatures have increased in all Arctic coastal areas examined. Changes in the Maximum Normalized Difference Vegetation Index (MaxNDVI) correlate both to changes in early summer coastal sea ice concentrations and summer land temperatures (Figure 3.21). In tundra regions, the annual maximum NDVI (MaxNDVI) usually occurs in early August and is correlated with aboveground biomass, gross ecosystem production, $CO_2$ fluxes, and numerous other biophysical properties of tundra vegetation. The data show that MaxNDVI has increased during the period of satellite observations (1982-2010) in Eurasia and North America (Dinniman et al., 2011), supporting model predictions that primary production of Arctic tundra ecosystems will respond positively to increased summer warmth (Bhatt et al., 2010; Lawrence et al., 2008).

Changes in temperature and vegetation, in turn, affect other ecosystems, including insects, birds, and animals. A 30-year observation of black guillemot bird colonies on an Arctic barrier island became witness to climate change impacts. The black guillemot, a diving seabird, has experienced major decreases in breeding success in arctic Alaska in the last decade due to the rapid and extensive retreat of summer sea ice. Formerly dependent on Arctic cod, the primary forage fish associated with arctic sea ice; guillemots now feed their nestlings lower quality and less abundant bottom-dwelling prey, resulting in lower breeding success and chick growth rates. Summer ice retreat has also shifted the range of the horned puffin, a nest competitor, and the polar bear, a nestling predator, further increasing guillemot nestling mortality (Moline et al., 2008; Smith et al., 2010). Changing habitat through loss of sea ice is threatening populations of polar bears in some locations who now must live more generally onshore during the summer and as a result are beginning to mate with grizzly bears (Kelly et al., 2010). Not all impacts to the ecosystem from climate change will necessarily be negative; for example, upper trophic levels may benefit and some organisms may experience extended ranges.

In Alaska's boreal forests, recent warming has been accompanied by reduced growth of formerly dominant

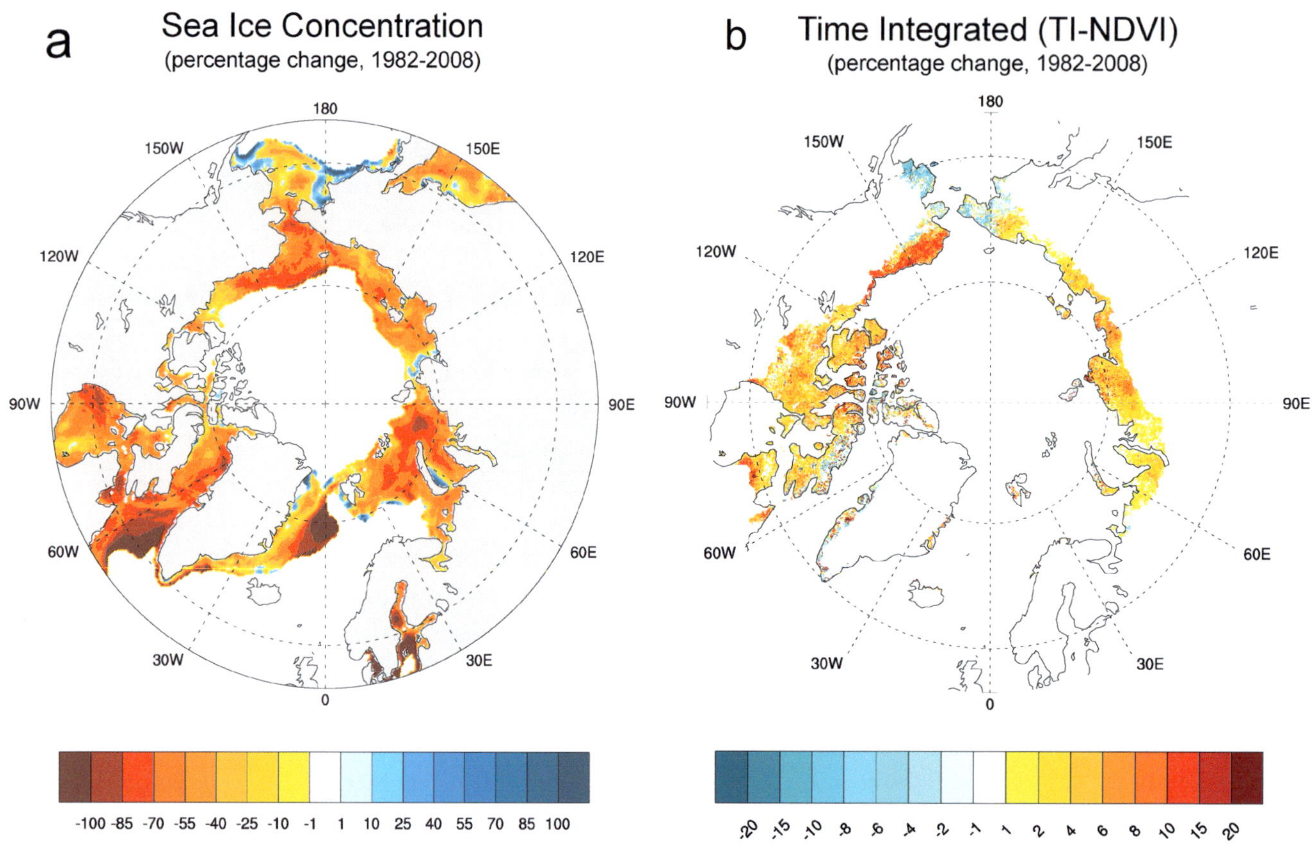


FIGURE 3.21 Percentage change in the Arctic from 1982 to 2008 (change in the 27-year trend expressed as a percent of the 1982 value) for (a) sea ice concentration at the 50 percent climatological value and (b) time-integrated Normalized Difference Vegetation Index (NDVI), showing that changes in the Arctic tundra have been correlated with increased temperatures in the Arctic Ocean. The data show that MaxNDVI has increased during the period of satellite observations (1982-2010) in Eurasia and North America (Dinniman et al., 2011), supporting model predictions that primary production of Arctic tundra ecosystems will respond positively to increased summer warmth. SOURCE: Copyright 2010 American Meteorological Society (AMS). Bhatt et al., 2010.

tree species (Calef, 2010; Juday et al., 2003), plant disease and insect outbreaks (IASC, 2011); drying of lakes (Roach et al., 2011); increased wildfire frequency, extent, and severity (Mack et al., 2011; Turetsky et al., 2010); and reduced safety of hunters traveling on river ice (Chapin and Lovecraft, 2011). In studies of the resilience of Alaska's boreal forest, Chapin et al. (2010) found that with continued warming, Alaska's boreal forest will undergo significant functional and structural changes within the next few decades that are unprecedented in the last 6,000 years. To look back over these time scales, one needs the perspective of paleoclimate studies.

## EVIDENCE OF PAST CLIMATE CHANGE OVER GEOLOGIC TIME SCALES

Knowledge of the past provides a context for the present. The precipitous minimum in Arctic summer sea ice extent in 2007 (Stroeve et al., 2007) gave global attention to the anomalous decline in sea ice since modern observations began, but it can only be understood fully when comparisons are made with data from long-term observations, or paleoclimate data. IPY provided a platform for a large number of national and international programs to exploit the essential temporal perspective provided by studies of past climate change at both poles.

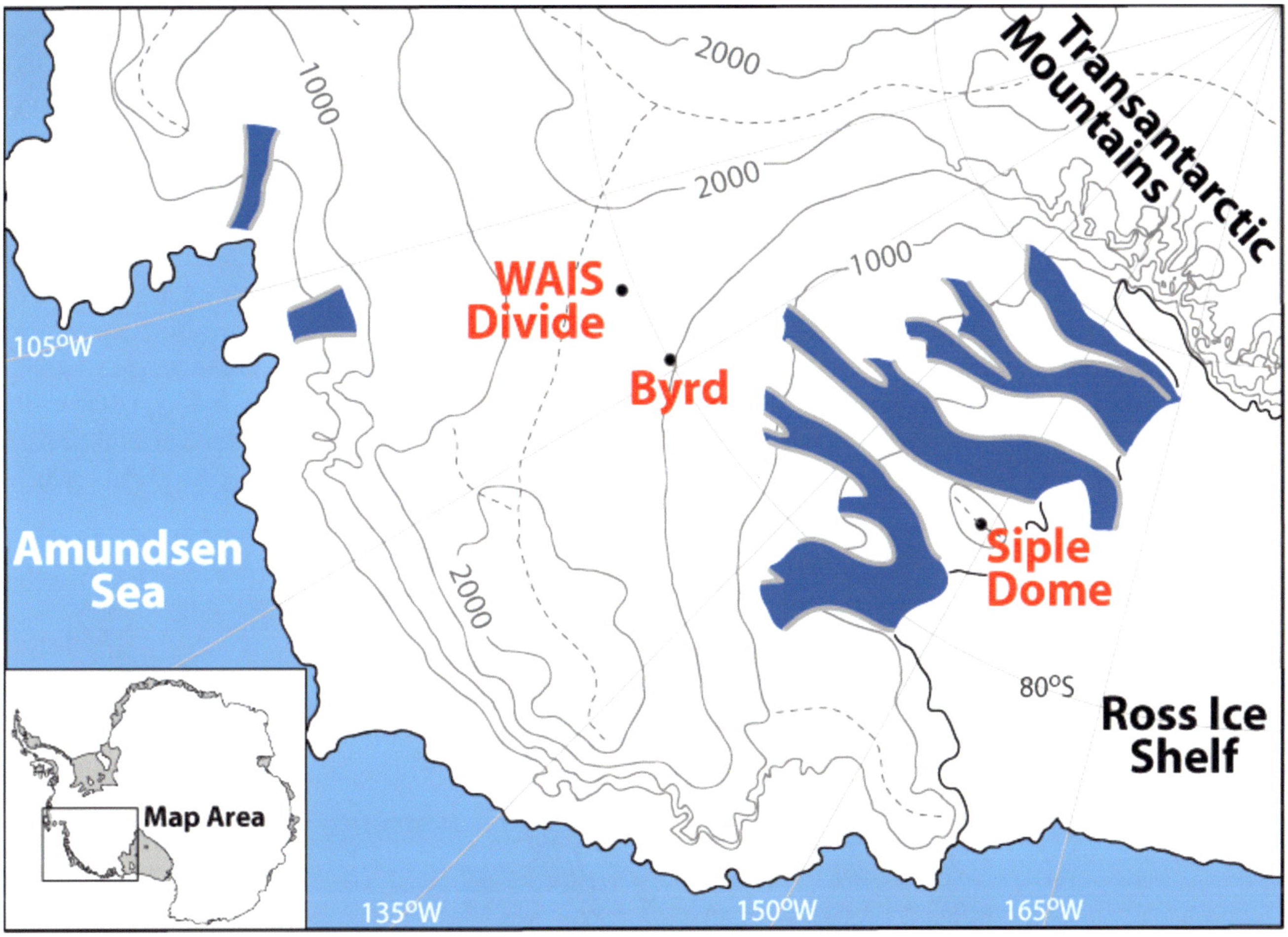


FIGURE 3.22 Location of the WAIS Divide Ice Core, which will be drilled at a cold, high-accumulation location on the West Antarctic ice sheet. The blue areas indicate fast-moving outlet glaciers and ice streams. SOURCE: Howard Conway, University of Washington.

### Ice Core Research

Ice core research was already under way during IGY (Langway, 2008), and it continued in this IPY to provide some of the most robust data concerning the climate history of our planet and of the sensitivity of the polar regions to warming (Alley et al., 2010). Drilling of the WAIS Divide (Figure 3.22) deep ice core in central West Antarctica began during IPY. Among other contributions, this core is poised to improve our understanding of rapid climate change events during the past glacial cycle, including impact of greenhouse gases. The recovery of a new 100,000-year-long ice core record from one of the most vulnerable portions of the Antarctic continent is producing the first Southern Hemisphere climate and greenhouse gas record with the same time resolution and duration as the ice core records from the Greenland ice sheet.

Early findings include the demonstration of how changes in tropical weather conditions in the tropics propagate into the Antarctic, highlighting the tight but distant linkages between currently occurring changes in climate (Ding et al., 2011). Steig et al. (2009) compiled up-to-date satellite observations and in-situ weather station data and found that most of West Antarctica has been warming significantly, particularly in winter and spring, since at least the 1950s (Vaughan et al., 2003). Two other recent studies also agree that the WAIS Divide site has been warming significantly (O'Donnell et al., 2010; Schneider et al., 2011). Preliminary borehole thermometry work[10] confirms this independently. Analysis of global sea surface temperature data and atmospheric reanalysis data products has shown that the

[10] Anais Orsi and Jeffrey Severinghaus, Scripps Institution of Oceanography, personal communications, 2011.

warming at WAIS Divide is a response to warmer sea surface temperatures in the central tropical Pacific Ocean, primarily associated with several large El Niño events in the 1990s (Ding et al., 2011; Schneider et al., 2011). The atmospheric circulation changes associated with these events appear to have been responsible for the significant increase in the inflow of warm circumpolar deep water onto the Antarctic continental shelf, resulting in increased thinning of ice shelves in the Amundsen Sea sector (Steig et al., 2011).

The IPY NEEM (North Greenland Eemian Ice Drilling) international ice core project (Figure 3.23) on Greenland (77.45°N 51.06°W), also drilled during IPY, had the ultimate goal of recovering a continuous ice record representing the last interglacial: the late Pleistocene Eemian period from 131,000 to 114,000 years ago. The purpose of this program was to better define the extent to which the Greenland ice sheet melted during this period. Greenland temperatures are thought to have been about 3-5°C warmer than present during the Eemian due to differences in the shape of Earth's orbit around the sun, making the Eemian time period a useful analogue for future climate warming caused by anthropogenic increases in $CO_2$ (Alley et al., 2010; Miller et al., 2010b; Otto-Bliesner et al., 2006). Moreover, this melt history is being used to evaluate

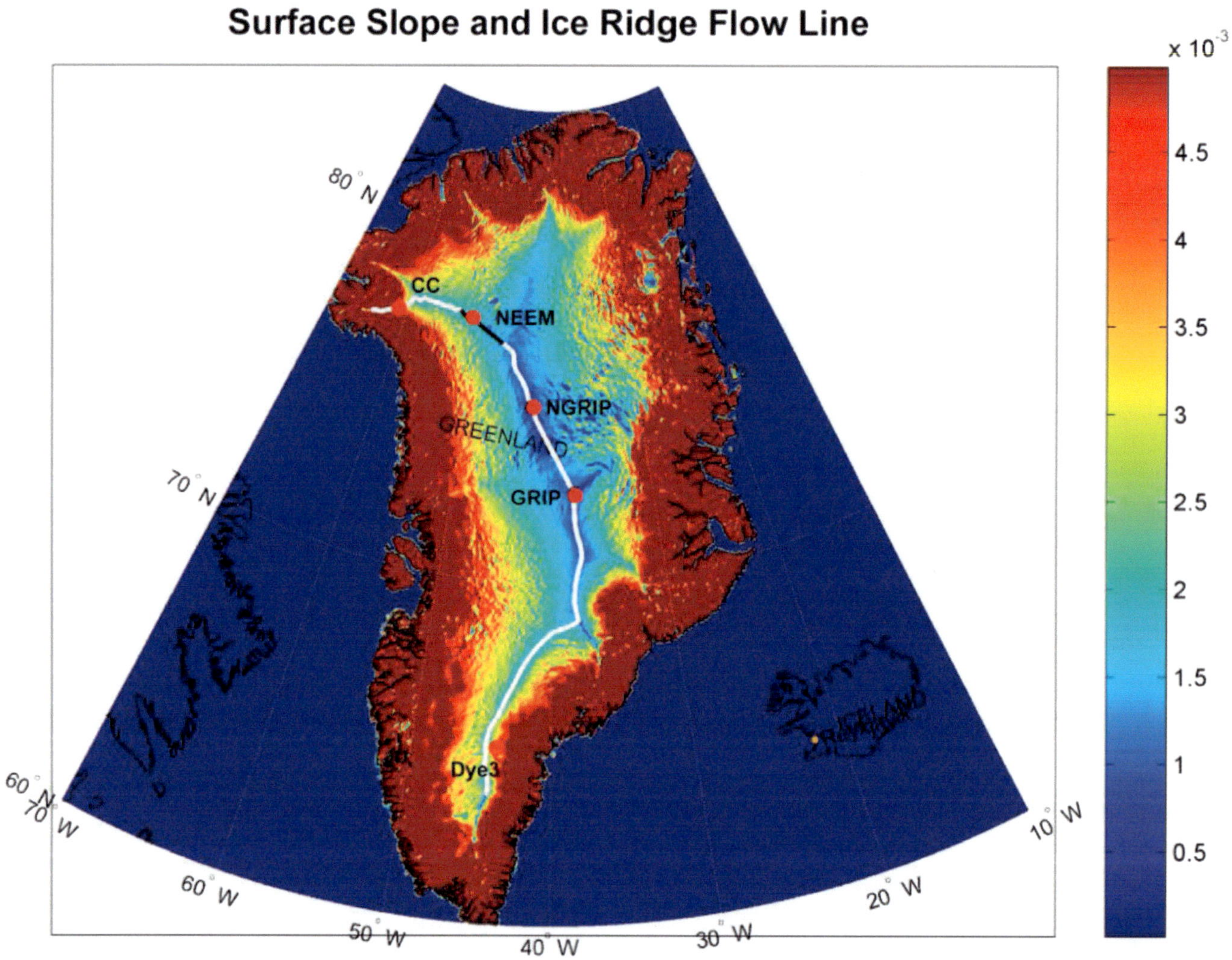


FIGURE 3.23 During IPY, the NEEM (North Greenland Eemian Ice Drilling) project had the goal of recovering a continuous ice record representing the last interglacial. Studying the amount of melt that occurred on the Greenland ice sheet during this period can help advance modeling efforts and enable scientists to understand future climate warming caused by anthropogenic increases in $CO_2$. This map shows the NGRIP (North Greenland Ice Core Project) and NEEM field camps in Greenland. The white and blue line connecting the two sites shows the approximate route followed by both the surface traverse and airborne survey in 2007. SOURCE: NEEM.

rapid climate change events in modeling efforts to better define the extent to which future anthropogenic $CO_2$ emissions will cause and/or accelerate loss of portions of the Greenland ice sheet (see Steffensen et al., 2008).

### Sediment Core Research

The new IPY ice cores are ideal for understanding high-resolution, rapid climate variability through the last glacial-interglacial cycle and their relationships with greenhouse gases. Sediment core records from the high latitudes obtained via deep geologic drilling during IPY, including the Antarctic Drilling Program (ANDRILL) and the International Continental Drilling Program (ICDP) Lake El'gygytgyn Project (Arctic Beringia), also provide extraordinary insights into older events in the complex climate evolution of the Arctic and Antarctic regions.

During the austral spring and summer of 2006-2007, ANDRILL recovered a 1,285-m-long sedimentary and rock core record of climate and ice sheet variability spanning the last 13 million years from beneath the McMurdo Ice Shelf (see also "Paleoclimate Tools" section in Chapter 4). The data included several climate cycles during the early Pliocene; this was a time when temperatures were ~3°C warmer than today and atmospheric $CO_2$ levels may have reached 400 ppm. The data imply a significant change in thermal regime of WAIS during the late Pliocene, coincident with global cooling in oxygen isotope records and the onset of Northern Hemisphere glaciations (Naish et al., 2009). Equally significant were innovative new approaches in numerical modeling that led to accurate simulations of marine-based ice sheet dynamics (Pollard and DeConto, 2009; Figure 3.24). These studies provided robust data/model validations for past fluctuations of the WAIS, including its total disappearance during the early Pliocene and during "super-interglacial" warmth of Marine Isotope Stage 31 (about 1.1 million years ago [Ma]; Scherer et al., 2008). Probably one of the most extraordinary outcomes of this ANDRILL research is that it demonstrated, for the first time, instability in the West Antarctic ice sheet, revealing Antarctica's vulnerability to oceanographic warming and changes in atmospheric $CO_2$, with dramatic consequences for future changes in global sea level. This important

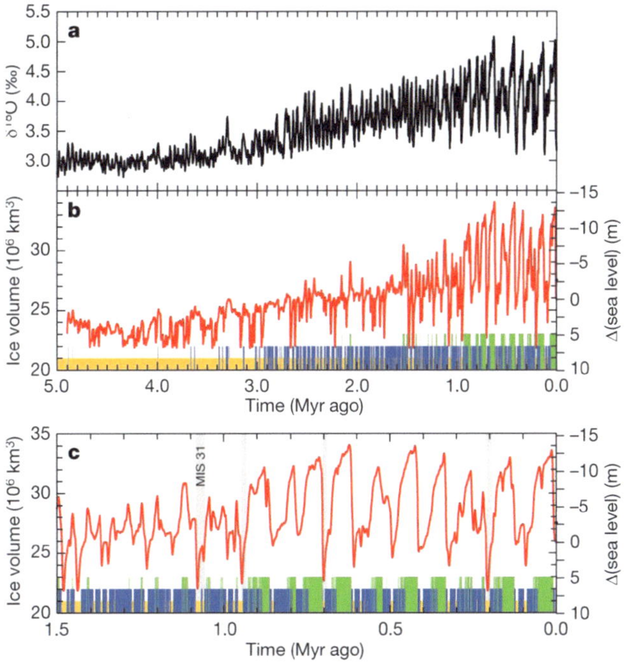


FIGURE 3.24 Sediment core records from the ANDRILL Program during IPY provide valuable information on climate evolution in the polar regions. For example, ANDRILL core records recovered from beneath the McMurdo Ice Shelf revealed data on climate cycles during the Pliocene. This ANDRILL model figure demonstrates past fluctuations and shows the loss of the WAIS repeatedly in Pliocene. Data from the cores and new approaches in numerical modeling of marine-based ice sheet dynamics can help us understand future changes in global sea level. SOURCE: Pollard and DeConto, 2009.

outcome was not revealed in earlier work (e.g., Denton et al., 1993; Marchant et al., 1993). Reconstructions of Antarctic paleogeography (Wilson and Luyendyk, 2009) have been critical for producing a better match between geologic data and modeled ice sheet response, suggesting possible locations of elevated topography for the initial buildup of ice on Antarctica as early as 34 Ma that were missing in earlier reconstructions.

During the austral spring of 2007-2008, ANDRILL's 1,138-m-long sedimentary core from the southern McMurdo Sound contained a near-continuous coastal record of Antarctic climate and ice sheet variability, including an interval of sustained global warmth known as the Mid-Miocene Climatic Optimum. In contrast to the dramatic shifts between glacial-interglacial end member climate states evident in Pliocene-Pleistocene and Oligocene drill cores on the same coastal margin, the early Miocene paleoclimatic fluctuations were subtler under warmer, more equable climate conditions (Harwood et al., 2009). Fossils preserved in these strata suggest marine climate conditions similar to that of southern Patagonia and southwestern New Zealand today, influenced by high sediment discharge from river runoff, and high coastal turbidity, implying surface air temperatures warm enough for significant ice surface melt and the transfer of moisture from the ocean onto the land and ice surface.

Comparable lengthy records of the Late Cenozoic history of the Arctic are poorly known. To partially fill this gap, during the boreal winter of 2008-2009, the science community successfully recovered a 3.6-million-year-long sediment record from Lake El'gygytgyn ("Lake E"), which is a 12-km-diameter meteor crater lake located 100 km north of the Arctic Circle in northeastern Russia (see also "Paleoclimate Tools" section in Chapter 4). The record captures for the first time the rhythm of orbitally forced climate in the Arctic. The Arctic climate was dominated by 41,000-year climate cycles up until about 1 million years before the present; since about 900,000 years ago, the Arctic has been dominated by climate cycles of 100,000 years, but with cycles of 21,000-23,000 years that persist in the precession band.

The Quaternary section of the Lake E sediment core includes a complete record of glacial-interglacial change, including warm intervals correlative with well-known marine isotopic stages. The extent to which many of these interglacials, including marine isotopic stages (MIS) 9, 11, and 31 and others, appear to have been warmer than MIS 5e (Lozhkin and Anderson, 2011), is an extraordinary surprise because it suggests repeated intervals in the past when the Greenland ice sheet may have been much smaller than today and sea ice vastly reduced (Melles et al., 2011). These new data contribute to numerical modeling

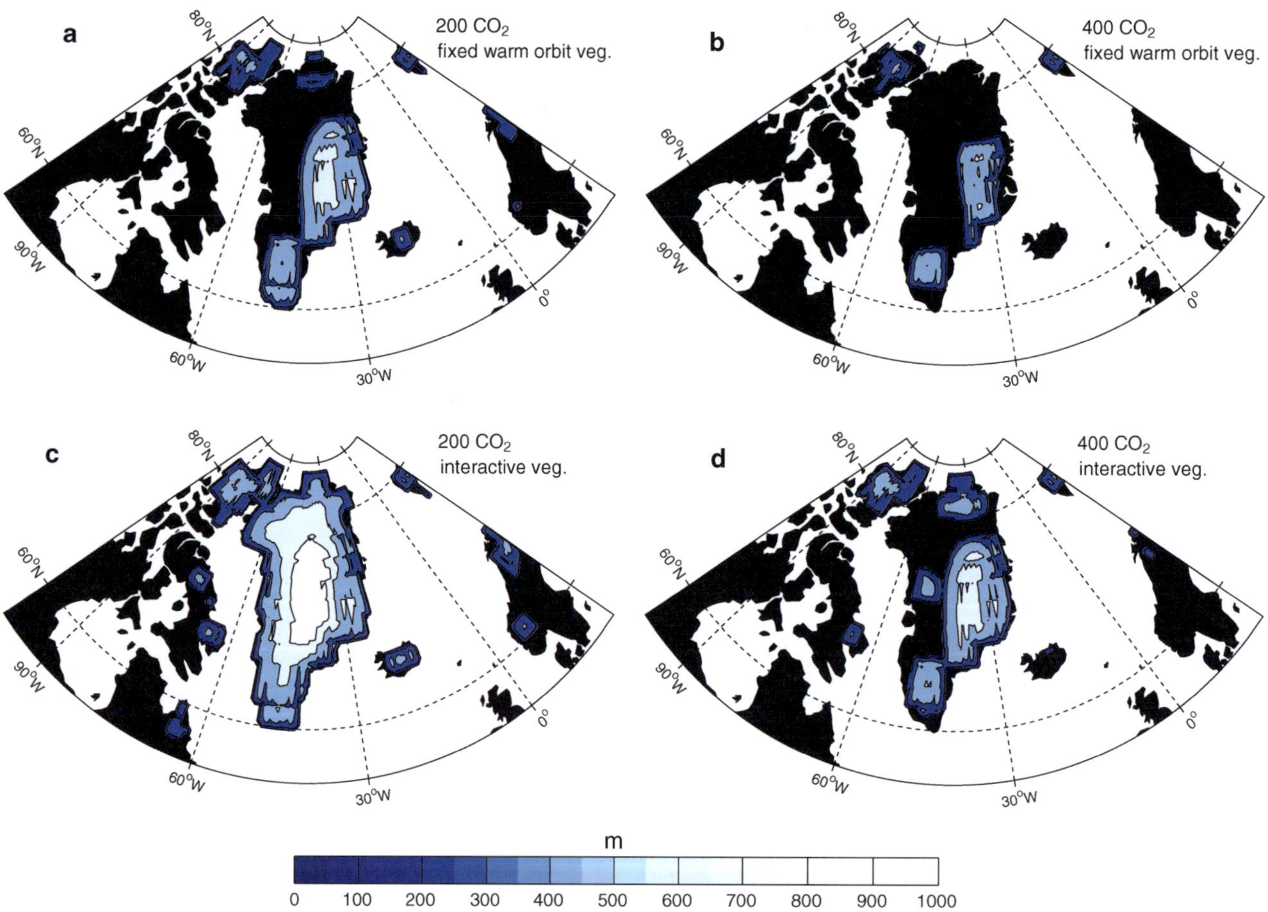


FIGURE 3.25 Simulation of inception of ice sheets on ice-free isostatically rebounded Greenland as simulated by the 3-D dynamical ice sheet model and forced by the sensitivity scenarios. Ice sheet thicknesses (m) are shown after 11,000 years for Pleistocene 200 ppmv $pCO_2$ (left panels) and Pliocene 400 ppmv $pCO_2$ (right panels) scenarios. (a, b) Cold orbit, fixed warm orbit vegetation; (c, d) cold orbit, interactive vegetation. SOURCE: Koenig et al., 2011.

efforts that test the vulnerability of Arctic sea ice and the Greenland ice sheet to global air temperature rise (Koenig et al., 2011; Figure 3.25). Extreme warmth at 1.1 million years ago during Marine isotope stage 31 is especially interesting because this interval coincides by half a precession cycle with the last time ANDRILL chronicles the collapse of the WAIS and the Ross Ice Shelf (Naish et al., 2009). The climate record from Lake E, especially the history of past interglacials, provides a means of testing what controls polar amplification over time using data/model comparisons.

## Paleoclimate Synthesis During IPY

For its part, the U.S. Climate Change Science Program commissioned a synthesis of paleoclimate data from the Arctic as one of many scientific synthesis reports intended to inform public debates on modern climate change (Fitzpatrick et al., 2010). The original report (CCSP, 2009) highlighted the record of climate change in the Arctic over the past 3-4 million years, in the context of a global system. This synthesis acknowledged that the anomalous loss of summer sea ice in 2007 is dramatic compared to recent glacial-interglacial cycles of natural variability but not unprecedented for warmer interglacial stages in the past (Polyak et al., 2010). At the same time, changes in the size of the Greenland ice sheet over its history have been primarily controlled by temperature, with warming in the past always causing considerable ice sheet shrinkage contributing to sea level rise (Alley et al., 2010). Arctic amplification, as a process in modern records of global climate change is now known to be a persistent feature of past climates, typically falling in the range of 3-4°C over global mean temperatures (Miller et al., 2010a; Figure 3.26). Paleoclimate ice core records show that once thresholds in the system are passed, global climate change can be larger and faster than models used now for predictions of future change might predict (White et al., 2010).

During IPY, multiple paleoclimate studies were carried out, aimed at characterizing natural climate variability, especially warm climate variability, and the system processes that drive variability on annual to millennial time scales. A community compilation of lake sediment sequences, ice cores, and tree ring records from the circumarctic region (Figure 3.27) confirmed

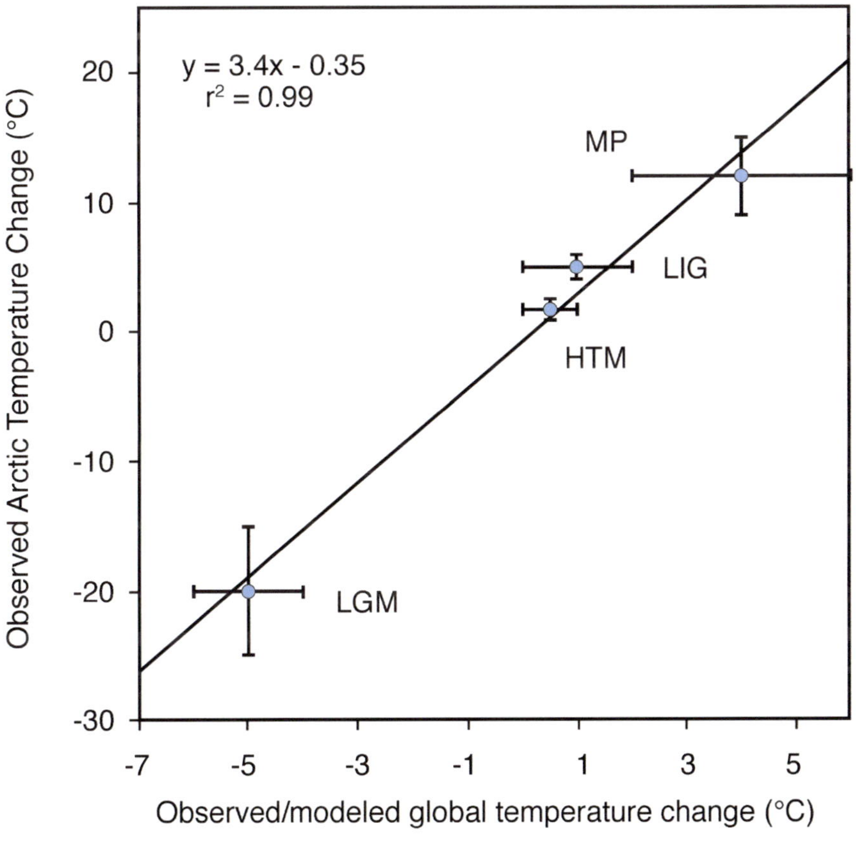


FIGURE 3.26 Paleoclimate data quantify the magnitude of Arctic amplification. Shown are paleoclimate estimates of Arctic summer temperature anomalies relative to recent, and the appropriate Northern Hemisphere or global summer temperature anomalies, together with their uncertainties, for the following: the last glacial maximum (LGM; ~20 ka), Holocene thermal maximum (HTM; ~8 ka), last interglaciation (LIG; 130 to 125 ka) and middle Pliocene (~3.5 Ma). The trend line suggests shows that summer temperature changes are amplified 3 to 4 times in the Arctic. SOURCE: Miller et al., 2010a; White et al., 2010.

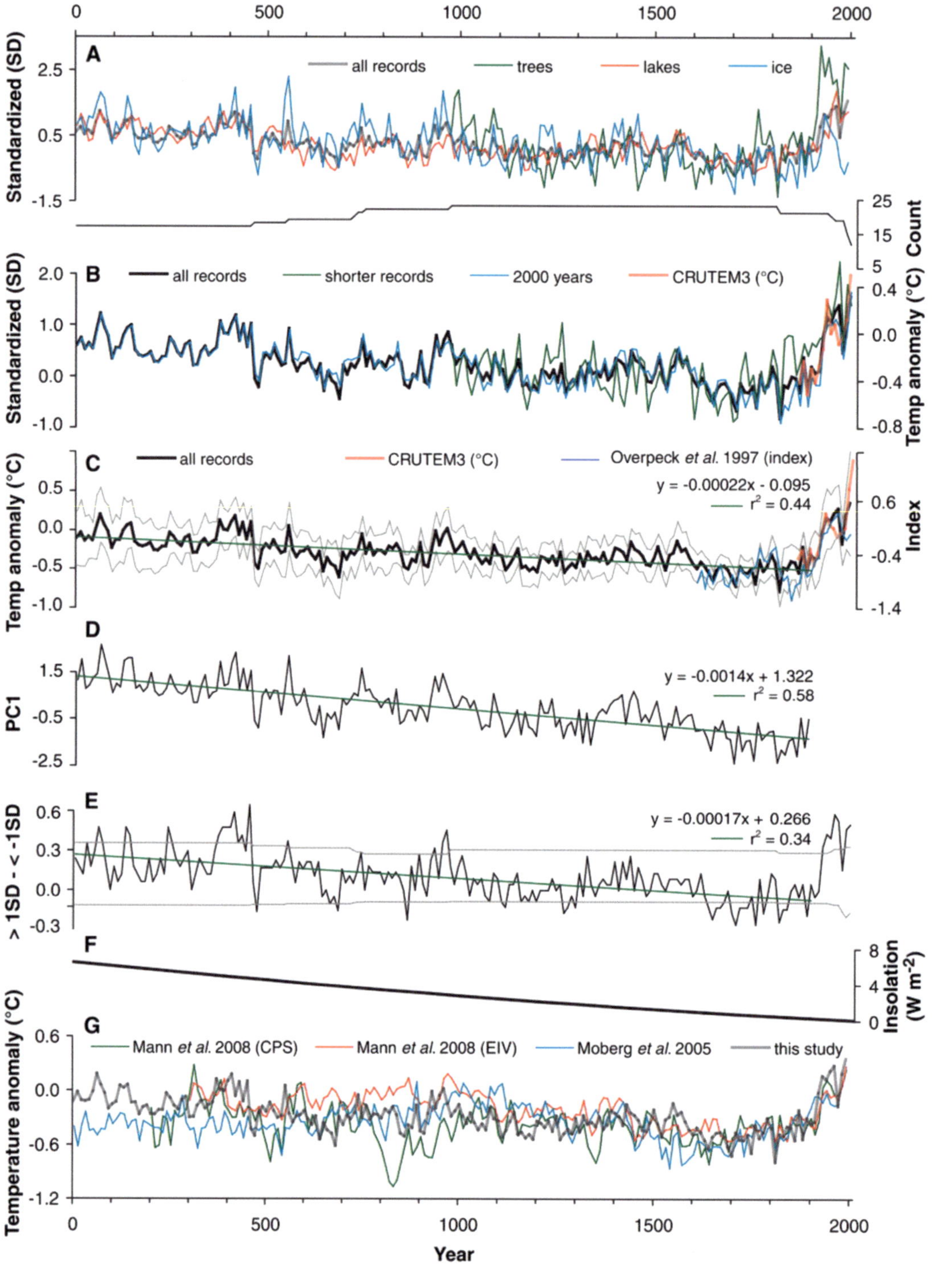


FIGURE 3.27 This community compilation of lake sediment sequences, ice cores, and tree ring records from the circumarctic region confirmed that the natural cooling trend of the last 2,000 years has been completely reversed by contemporary warming, with the last five decades being the warmest over the length of record. (A and B) Kaufman et al. (2009) show a composite of 23 high-resolution proxy climate records from the Arctic. (C) Mean of all records transformed to summer temperature anomaly relative to the 1961-1990 reference period, with first-order linear trend for all records through 1900 (green line), the 400-year-long Arctic-wide temperature index of Overpeck et al. (1997) (blue curve; 10-year means), and the 10-year-mean Arctic temperature through 2008 (red line). (D) Time series of PC1 based on the 15 records that extend from 1 C.E. to 1900 C.E., showing a strong first-order trend. (E) Difference in the fractional proportion of records that exceed ±1 SD for each 10-year interval. (F) Change in summer (JJA) insolation at 65°N latitude relative to the 20th century (Berger and Loutre, 1991). (G) Northern Hemisphere average proxy temperature anomalies (10-year means) reconstructed by Mann et al. (2008) on the basis of two approaches and by Moberg et al. (2008). The Arctic regional reconstruction is overlaid in gray. SOURCE: Kaufman et al., 2009.

that the natural cooling trend of the last 2,000 years has been completely reversed by contemporary warming, with the last five decades being the warmest over the length of record (Kaufman et al., 2009). Because the climate has shown a normal response to natural forcings over the last two millennia, the only recent forcing capable of causing this reversal is the dramatic input of anthropogenic carbon dioxide to the atmosphere. This clear attribution for anthropogenic forcing of current climate change came not only from a synthesis of the geologic and glaciological record, but also from the modeling community (Miller et al., 2010a; Serreze and Barry, 2011), as well as from the space physics community, who used satellite observations to show that energy from the sun was a minor impact in climate change (Scafetta and West, 2005).

Paleoclimate data also provide independent evidence for comparison and calibration of climate model simulations of past change. For example, the last time Arctic summer sea ice extent was vastly reduced was 6,000-8,500 years ago, when solar insolation was about 7 percent higher than today because Earth's orbit reached perihelion in northern hemisphere summers (Funder et al., 2011). Sea ice was also vastly reduced 125,000 years ago, when solar insolation was 11 percent more than now. But in both cases, atmospheric $CO_2$ levels were only 270-280 ppm (Miller et al., 2010a; Polyak et al., 2010). Assessments of the rate of climate change in the past show that climate responses can be rapid with diminishing sea ice first being a feedback (i.e., in this case through changes in albedo as ice transitions to open water) and eventually becoming a forcing (i.e., a primary driver of climate change) with delayed release of heat from areas ice free in summer (Serreze and Barry, 2011; White et al., 2010). Environmental change, which occurred on longer time scales in the past, is now happening faster than models predict. This is a major emerging theme of IPY 2007-2008.

Improved understanding of teleconnections between the Arctic and Antarctic and mid- to lower-latitude climate regimes are emerging from IPY paleoclimate studies. As is observed today, when $CO_2$ increased in the past, the global system warmed up with an amplified response across the polar regions, especially the Arctic. Changes in sea ice and the extent of ice sheets create feedbacks in the climate system with implications for regional change and sea level fluctuations around the world. While long-term models tend to underestimate polar amplification, projections from these paleoclimatic analogs point to the possibility that human influence will become unprecedented in combined speed and persistence (White et al., 2010).

## ARCTIC SOCIETIES AND SOCIAL PROCESSES

The IPY planners made a radical departure from the earlier IPY/IGY template by creating a special "People" field and by introducing the social sciences and humanities, as well as the studies of human health to the IPY program:

- Research Theme #6: To investigate the cultural, historical, and social processes that shape the sustainability of circumpolar human societies, and to identify their unique contributions to global cultural diversity and citizenship (Rapley and Bell, 2004).
- Observational Strategy #6: To investigate crucial facets of the human dimension of the polar regions which will lead to the creation of data sets on the changing conditions of circumpolar human societies (Rapley and Bell, 2004).

Previous IPY/IGY excluded research in the socio-economic and humanities fields, except for the limited medical studies carried on the personnel of the polar stations. This new and more human-oriented format of IPY reflected more integrative and society-driven nature of today's polar research. In the United States alone, NSF allocated almost $20M in support of more than 30 research, observational, and data management projects in the social sciences and the humanities, the largest-ever concerted U.S. funding for such efforts. IPY's new focus on the social and humanities issues was also spearheaded by the preceding efforts stimulated by the Arctic Council, such as the *Arctic Human Development Report* (AHDR, 2004), *Arctic Climate Impact Assessment* (ACIA, 2005), *Survey of the Living Conditions in the Arctic* (SliCA; Andersen et al., 2002) and others (Downie and Fenge, 2003) that were initiated prior to, or during, the planning phase for IPY. These and other new developments resulted in the increased engagement of polar residents, particularly Arctic indigenous people, in IPY operations. Such broadening of the research base and scope led to significant science breakthroughs in IPY and in the general conduct of modern research at the poles.

Altogether, IPY social science and humanities projects engaged more than 1,500 researchers, students, indigenous experts and monitors, and representatives of Arctic indigenous organizations in more than 30 international and numerous national research and outreach projects (Krupnik et al., 2011). They studied human societies, past and present, and sought better understanding of forces that govern social interactions during IPY; they developed new approaches, interpretive models, and groundbreaking research paradigms.

There have been many pioneer advances in the polar social sciences during the IPY era. Although most of the IPY social science and humanities efforts were locally focused, several international projects included new coordinated research and data collection in four or more Arctic nations. They produced the first-ever broad circumpolar overviews of local community adaptation and vulnerability (Hovelsrud and Smit, 2010), status of indigenous reindeer herders' and caribou hunters' knowledge (Oskal et al., 2009), indigenous use of the sea ice habitats (Huntington et al., 2010), role of governmental policies in community resettlement and relocations (Schweitzer et al., in press), and other research fields.

New "baseline" data sets were generated on community development, industrial exploitation of polar resources, status of indigenous languages and knowledge systems, cultural heritage, community well-being (Larsen et al., 2010), and the community use of local resources. IPY researchers have connected these data to the earlier datasets, including those built by previous statistical surveys (SliCA—Andersen et al., 2002), thus expanding the scope of IPY records by several decades (Hamilton, 2009; Heleniak, 2008, 2009; Kruse, 2010; Winther, 2010). Still, the geographic coverage of IPY activities in the social sciences and the humanities was quite uneven, with the bulk of research conducted in the Eastern Canadian Arctic (Nunavut, Nunavik), Alaska, Norway, and Greenland, and fewer efforts in the Russian Arctic, northern Finland, and Iceland. Even within better covered regions some communities received more attention, like Barrow, Gambell, Shismaref, Toksook Bay, and others in Alaska, while many more were hardly touched by IPY.

IPY-generated research introduced a new vantage point in assessing environmental change at the poles, namely, the stock of knowledge by local residents and, especially, polar indigenous people. It includes records of generation-based observations and extensive local terminologies of sea ice and snow patterns and phenomena, often of many dozen terms (Krupnik et al., 2010a,b; Oskal et al., 2009). Many scientists and indigenous experts now believe that the vantage points offered by "two ways of knowing" (Barber and Barber, 2007), academic research and local/indigenous knowledge, are needed for a comprehensive understanding of the polar regions and processes. The changes in polar sea ice, for example, are observed and assessed differently and at various scales by ice scientists, climate modelers, oceanographers, local subsistence users, and social scientists (Eicken, 2010; Eicken et al., 2009).

Even though the ultimate goals of scientists (understanding and modeling of climate change) and polar residents (sustainable adaptation) may be different, each group can learn from the vision of the others, and the common resulting knowledge is more than the sum of its individual parts. By adding a sociocultural perspective and indigenous knowledge (Box 3.2), scientists broadened the IPY agenda in sea ice research beyond its common focus on ice dynamics and coupled ocean-atmosphere-ice modeling. One of the key IPY legacies is the legitimization of these multiple "ways of knowing" (Huntington et al., 2007; Kofinas et al., 2010); it marked a revolutionary paradigm shift accomplished during the IPY era.

Additionally, prior to IPY the prevailing pattern of modeling complex linkages and impacts of climate change was to place "humans" at the bottom of the chain-like charts illustrating interconnections within the ecosystem. The underlying assumption was to explore how "humans" (i.e., people or communities) respond to the impacts projected by computer-generated scenarios, such as warming climate, shorter ice season, or thawing permafrost. The new approach explored during IPY, called community-based vulnerability assessment, has moved communities to the center of the study of climate change (Hovelsrud and Smit, 2010). It starts with the observations of change reported within local communities and by their members and it proceeds bottom-up to identify potential new conditions,

**BOX 3.2**
**Multiple Ways of Understanding Sea Ice**

Ice scientists, climate modelers, oceanographers, local subsistence users, anthropologists, mariners, and science historians have remarkably different visions of polar sea ice. To various groups of scientists, sea ice is a multifaceted physical and natural entity—an ocean-atmosphere heat flux regulator, a climate trigger and indicator, a habitat (platform) for ice-associated species, and/or an ecosystem built around periodically frozen saltwater. To polar explorers and historians, sea ice was first and foremost a formidable obstacle to humanity's advance to the Poles (Bravo, 2010). Polar indigenous people view sea ice primarily as a cultural landscape, an interactive social environment that is created and re-created every year by the power of their cultural knowledge. It incorporates local ice terminologies and classifications, ice-built trails and routes with associated place names, stories, teachings, safety rules, historic narratives, as well as core empirical and spiritual connections that polar people maintain with the natural world (Krupnik et al., 2010a).

Cultural landscapes created around polar sea ice (icescapes) are remarkably long-term phenomena, often of several hundred years (Aporta, 2009). By adding a sociocultural perspective and indigenous knowledge, ice scientists broadened the IPY agenda in sea ice research beyond its habitual focus on ice dynamics and coupled ocean-atmosphere-ice modeling (Druckenmiller et al., 2010; Eicken, 2010).

opportunities, or risks that communities are facing or may face in the future. This approach includes many more parameters of change, both physical and sociocultural, such as local demographic and economic factors, migration patterns, support networks, educational level, and others (Hamilton et al., 2010; Huntington et al., 2007). It puts critical emphasis on the assessment of community responses to future risks, sensitivities and adaptive strategies, and it requires extensive data collection at the community level, as the current adaptation mechanisms are researched and understood.

Often more immediate challenges stem from the many social agents, such as local system of governance, economic development, breakup in community support networks, availability of health care, and culture shifts. In certain areas in the Arctic, the purported "threat" of climate change masks the impact of more immediate factors, such as the alienation of property rights, appropriation of land, disempowerment of indigenous communities, and more restricted resource management regimes (Forbes et al., 2009; Konstantinov, 2010). Not all impacts from climate change will necessarily be negative, and climate and broader environmental change and its many impacts should be thus viewed as an added stressor to the already challenging local conditions on the ground.

Two international IPY projects—Community Adaptation and Vulnerability in the Arctic Region (CAVIAR) and Arctic Social Indicators (ASI)—were particularly instrumental to this transformation. The CAVIAR project tested new research and modeling approaches to assess the vulnerability and adaptability of 26 local communities in Canada, the United States (Alaska), Greenland, Iceland, Norway, Sweden, Finland, and Russia (Hovelsrud and Smit, 2010). The main outcome was a new vision of the Arctic peoples' resilience to environmental stress as a "two-way" process that depends as much (or more) on the strength of the community internal networks (social, cultural, institutional, economic, etc.) as on the intensity of the environmental signals. The ASI project initiated by the Arctic Council developed a set of thoroughly calibrated indicators to evaluate the status of sociocultural well-being of Arctic population at the community, local, and regional levels (Crate et al., 2010). The previously used general national indexes used by UNESCO and other major international agencies, such as per capita gross domestic product or the overall level of literacy,[11] have been successfully superceded by more locally nuanced tools to assess community well-being as a result of IPY research.

Another critical frontier theme explored in IPY was the relationship between indigenous perspectives developed via generations of shared knowledge and the data and interpretations generated through scholarly research. The field that compares such perspective did not even exist prior to the late 1990s. Several IPY projects contributed to our increased understanding of how indigenous knowledge could be matched with instrumental data in monitoring the changes in Arctic ice (see Box 3.2; Figure 3.28), snow and vegetation condition,[12] marine mammal and caribou/reindeer migrations, and behavioral patterns of polar animals and fishes (Hovelsrud et al., 2011; Kofinas et al., 2010;

[11] *http://unstats.un.org/unsd/demographic/products/socind/default.htm.*

[12] *http://icr.arcticportal.org/index.php?option=com_hwdvideoshare&task=viewvideo&Itemid=127&video_id=11&lang=en.*

FIGURE 3.28 The "high resolution" of indigenous terms for sea ice often allows distinction among numerous types of ice and related phenomena within a small area. On this photograph, a female walrus and a calf (isavgalik) are resting on the ice (nunavat) in the midst of scattered pack ice (tamalaaniqtuaq) interspersed with patches of calm flat water (quuniq). The mass of floating ice (sigu) consists of various ice formations such as puktaat (large floes), puikaanit (vertical blocks of ice), kangiqluit (floes with overhanging shelves) taaglut (pieces of darker or dirt ice), and sangalait (small floating pieces of ice) SOURCE: Winton Weyapuk Jr., May 21, 2007; Krupnik and Weyapuk, 2010.

Krupnik and Hovelsrud, 2011; Oskal et al., 2009). Yet another "frontier" area advanced during IPY explores how to make polar research culturally and socially relevant to local residents. It argues for collaboration with the new groups of stakeholders on research planning in their home areas to assess local concerns and for the new research agenda to be set through dialogue with communities rather than via top-down planning by funding agencies or at university campuses.

Major outcomes from IPY social science and humanities research included the multilevel and adaptive nature of governance of the "international spaces," such as Antarctica, the Central Arctic Basin, High Seas and Outer Space (Berkman et al., 2011; Shadian and Tennberg, 2009). This outcome originated in large part from the extensive historical studies of IGY 1957-1958 and previous IPYs (Barr and Luedecke, 2010; Elzinga, 2009; Launius et al., 2010), of the implementation of the Antarctic Treaty of 1959 and the new role of the United Nations Convention on the Law of the Sea (UNCLOS) in the Arctic policy debate. Another frontier area pioneered in IPY was the comparative study of Northern-Southern Hemisphere processes under the concept of "fringe environments" (Hacquebord and Avango, 2009). In the social sciences and humanities fields, it focused on the history of polar explorations, commercial use of local resources, polar governance, tourism, and heritage preservation (Avango et al., 2011; Barr and Chaplin, 2008; Broadbent, 2009; Hacquebord and Avango, 2009); it illustrated remarkable parallels in human advances into both northern and southern polar regions.

In the years prior to IPY, the dichotomy between the northern and southern regions went far beyond the basic biological and physical differences exemplified by the northern polar bear and the southern penguin, or ocean ringed by continents in the north and continent surrounded by ocean in the south. Antarctic social research was almost nonexistent, as there were "no people" in Antarctica. As a result of IPY, a new network of the "Antarctic social sciences" emerged, first, in the form of SCAR Action Group (AG) on the History of Institutionalization of Antarctic Research (established in 2004 and focused primarily on the history of human explorations in Antarctica),[13] followed by another and much broader SCAR Social Science AG, "Values in Antarctica. Human Connections to the Continent," that includes specialists in political sciences, cultural and human geography, law, economics, tourism, literature, psychology, and media studies. [14] These developments were triggered by an explosion of interest in social issues that are common to both polar regions and

[13] *http://www.scar.org/about/history/.*
[14] *http://www.scar.org/researchgroups/via/.*

gaining speed in the post-IPY era (Liggett and Steel, 2011), including the Montreal "Knowledge to Action" Conference (April 2012).[15]

Overall, new research in the social science and humanities fields helped advance a broad variety of themes: the well-being of polar communities; use of natural resources and economic development, particularly the impact of oil and gas industry in the polar regions; local ecological knowledge; preservation of natural, historical, and cultural heritage and the status of indigenous languages; and history of exploration, peopling, and the exploitation of polar regions. IPY participants from local communities and polar indigenous organizations were particularly active in studies investigating adaptations to rapid environmental and socioeconomic changes. They joined forces with the IPY monitoring efforts to collect, exchange, and document data on sea ice, biota, and climate, use of local resources, and impacts of industrial exploitation of the polar regions (Hovelsrud and Smit, 2010). These and other contributions of polar residents to the IPY program make one of its most lasting achievements.

Besides innovative projects in social sciences, IPY also featured numerous activities in the humanities, both international and on the national and regional scales. Altogether, more than 20 international projects in the humanities were endorsed for the IPY program, including several museum exhibits ("Thin Ice," "Inuit Voices," "Antarctic Touring Exhibit," and others), numerous arts and media shows, books, and films (see examples in Kaiser [2010], Zicus [2011], and Chapter 5). Notable U.S. events include the FREEZE[16] activities in Anchorage in January 2009 celebrating Alaska and life in the North, where artists, architects, and designers from Alaska and around the world came together to create large-scale outdoor installations in downtown Anchorage using snow, ice, and light—distinctly northern elements. Further descriptions of activities that helped bring the IPY message to thousands of people worldwide are included in Chapter 5 ("Knowledge to Action").

## HUMAN HEALTH

IPY 2007-2008 was the first IPY to include human health dimensions as a recognized thematic area of study. IPY activities related to human health were primarily focused on the permanent inhabitants of polar regions, with additional efforts to reach transient and nomadic communities as well. Research previous to IPY had highlighted several discrepancies in basic health metrics between indigenous and nonindigenous populations residing in these areas (Young and Bjerregaard, 2008). Mortality proxies such as life expectancy at birth and infant mortality are generally less favorable (lower life expectancy at birth and higher infant mortality) for indigenous populations throughout the circumpolar world, though distinct regional differences persist. As such, it was recognized that IPY represented a unique opportunity to further stimulate cooperation and coordination on Arctic health research and end-user access.

The Arctic Human Health Initiative (AHHI[17]) was created during IPY to link researchers with potential international collaborators and to serve as a focal point for human health activities (described in section on "Subsistence Communities in the Arctic" in Chapter 5). While various networks exist to coordinate circumpolar health researchers, these projects exist on a widely variable country-by-country basis. One of the goals of AHHI was to enhance these systems, add international connectivity, and provide a better access to data resources. In the United States, the University of Alaska, Anchorage has established a new graduate program aimed at circumpolar health issues. In an effort to connect polar regions, the International Union for Circumpolar Health (IUCH[18]) now serves as an ongoing network where the many circumpolar societies can meet and work on initiatives that support research, development, networking, and dissemination of health information, including the Congress on Circumpolar Health, held every 3 years. IPY also saw the establishment of the International Network of Circumpolar Health Research, as well as ArcticNet, which serve to connect researchers from across the globe. To ensure that data access was made available to the many constituents of the project, a key focus of IPY was to create a legacy of data resources. Thus a human

[15] *http://apecs-social-sciences.blogspot.com/2011/08/fwd-ipy-montreal-22-27-april-2012.html.*

[16] *http://freezeproject.org/alaska/.*

[17] *www.arctichealth.org/ahhi/.*

[18] *http://iuch.net/.*

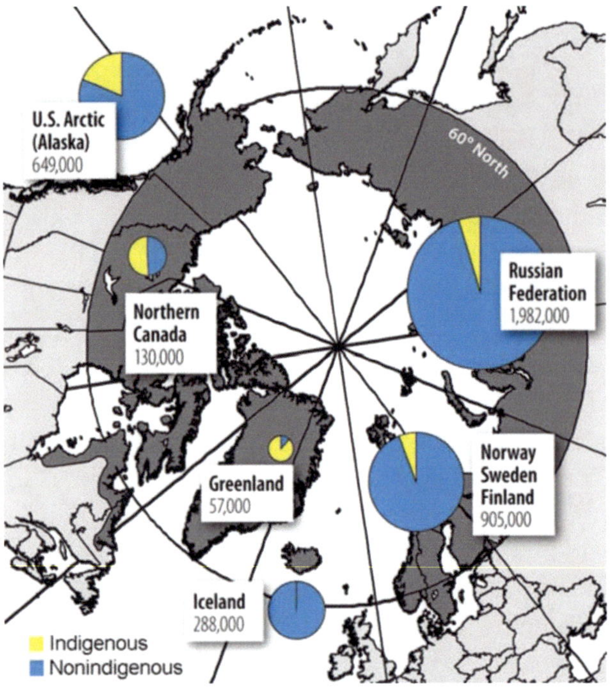


FIGURE 3.29 The ICS network of public health laboratories and institutes collects, compares, and shares data on infectious diseases. This map indicates participating countries (areas shaded dark gray), as well as the locations of clinical laboratories (small and large dots) that were used to monitor cases of invasive disease. During IPY, efforts were made to incorporate this type of data into SAON. SOURCE: Parkinson et al., 2008.

health component was incorporated into the Sustaining Arctic Observing Networks (SAON) that pools existing networks, such as the infectious-disease-oriented International Circumpolar Surveillance (ICS) system (Parkinson et al., 2008; Figure 3.29), to form a central site for human health-related concerns. These projects are deemed essential not only to the research mission of IPY, but to ensure that user needs are incorporated and prioritized on many levels.

IPY health research focused on a suite of issues of concern to Arctic residents, including health impacts of environmental contaminants, climate change, rapidly changing social and economic parameters within communities, chronic diseases, and health disparities between indigenous and nonindigenous residents.

Research on environmental contaminants was targeted at understanding how modern pollution affects indigenous life. Though socioeconomic circumstances and lifestyle contribute to health determination, studies also show that contaminant levels in some parts of the Arctic have the potential for adverse health effects. Epidemiological studies have related immunological, cardiovascular, and reproductive effects due to contaminants present in some Arctic populations (AMAP, 2009). Another study[19] is examining the risk of breast cancer in Inuit women in response to POPs, and further studies are investigating whether climate change contributes to high levels of POPs in fish and humans. Results of these studies are forthcoming and should provide insight into the relationship between native populations and their environment in a changing world.

Another major health discrepancy is the rate of infectious diseases seen among indigenous populations as compared to nonindigenous. In response to the high native rates of hepatitis B infection, for example, member nations have established the Circumpolar Viral Hepatitis Working Group and are conducting studies to determine the epidemiology.[20] Already, this group has identified a new HBV subgenotype (B6), unique to some native populations (Sakamoto et al., 2007), and has investigated outbreaks within the community (Børresen et al., 2010). Sexually transmitted infections are also high within indigenous populations (Gesink-Law et al., 2008) and work identifying at risk communities has shown that social and cultural norms significantly impact this problem (Gesink-Law et al., 2010; Rink et al., 2009). Beyond these activities, studies looking at the prevalence of zoonoses and parasitic infections (Gauthier et al., 2010), *Streptococcus pneumoniae* (Bruce et al., 2008), and human papillomavirus,[21] have addressed serious health issues in native populations and provide a research base for studies looking forward.

Because lifestyle changes have engendered an increase in obesity, diabetes, and cardiovascular disease in native populations (Galloway et al., 2010), several IPY studies were chartered to address these issues. "The Inuit Health in Transition" was a large international study focused on diet and lifestyle factors (smoking, physical activity, etc.) and is currently tracking living conditions, lifestyle risk factors, and environment with their relationship to chronic disease (Chan et al., 2009; Dewailly et al., 2009). Preliminary finds are now being distributed. Similar studies at University of Alaska, Fairbanks are building collaborative research presences

[19] *http://classic.ipy.org/development/eoi/details.php?id=1257.*
[20] *http://classic.ipy.org/development/eoi/details.php?id=1109.*
[21] *http://classic.ipy.org/development/eoi/details.php?id=1121.*

FIGURE 3.30 During IPY, there were a number of initiatives to explore behavioral and mental health issues in the northern regions. For example, the Inuit Health Survey team visited 36 communities during the summers of 2007 and 2008 to collect information on mental and community wellness. Locations of the Inuit Health Survey are shown on this map. SOURCE: Steven Fick/Canadian Geographic.

in Native communities focusing on the reduction of health disparities (Mohatt et al., 2007).

Beyond addressing physical illness, depression and suicide have been highlighted as significant issues in northern regions (Levintova et al., 2010). During IPY, there were a number of research projects that explored behavioral and mental health issues and the relationships between outcomes and environmental factors. The Inuit Health Survey[22] (Figure 3.30) collected information on mental and community wellness and provided information on their prevalence and evaluated community support and other determinants of resilience (Egeland, 2009). A Nunavik cohort study focused on the exposure of environmental contaminants and lifestyle factors (smoking, drugs, alcohol) on child behavior and development (Muckle et al., 2009). Studies looking at adoption (Laubjerg and Petersson, 2010), culturally based preventive intervention (Allen et al., 2009), and rapid social transition[23] also have tackled some of the problems unique to this demographic.

Circumpolar regions experience unique challenges in the delivery of health services because of the dispersed populations and geographic isolation. In response to this, the Northern Forum (NF[24]) was established to promote mutually beneficial collaboration in telemedicine, telehealth, mobile medicine and distance learning. This project is an important first step in both improving technologies and enhancing forums to promote partnership activities. Beyond this involvement, offerings including numerous symposia and workshops, published books and journals,

[22] *www.inuithealthsurvey.ca/?nav=home.*

[23] *http://classic.ipy.org/development/eoi/details.php?id=1266.*

[24] *www.northernforum.org/.*

television and radio presentations, and establishment of educational programs have all enhanced the access and connection to IPY activities.

In conclusion, this was the first International Polar Year to address Arctic health issues, and first results are still emerging. By establishing the infrastructure, connectivity, and dissemination products and prioritizing them around user needs, a system has been in place to provide support for this research mission and user interface for years to come. This is an important new direction in science that is a distinct and important legacy of IPY.

## CONCLUSIONS

Scientific discoveries during IPY used observations from some of the most remote regions of the Earth for a new understanding that benefits all humanity. Clear attribution that current warming of the planet is due to human activity came during IPY from at least three totally different research areas, the paleoclimatology, space physics, and modeling communities. Lake sediment sequences, ice cores, and tree ring records from the circumarctic show that recent warming has reversed the cooling trend of the last 2,000 years. Warming and freshening of the Arctic Basin is increasing, having a large impact on both sea ice reduction and basin stratification. The changes are having significant impacts at all trophic levels of the marine environment—from microorganisms to top predators in both polar regions. Terrestrial research show that land warming, sea ice decline, and greening of the Arctic are linked; this observation of modern processes is supported by paleoclimate findings on terrestrial systems. A new realization emerged that the total belowground carbon pool in permafrost is more than double the atmospheric carbon pool and three times larger than the total global forest biomass; this potentially provides an additional positive feedback parameter in the global system.

Discoveries involving the mechanisms of ice sheet flow associated with internal hydrological and subglacial conditions and interaction of ice shelves with the warming ocean enabled new understanding of ice sheet stability. The West Antarctic ice sheet became unstable and collapsed repeatedly, significantly raising sea level, during the interglacials of the past 3.5M years, which were warmer than today. Paleoclimate data show repeated intervals in the past when the Greenland ice sheet may have been much smaller than today and sea ice reduced. The IPY years spawned the realization that the impacts of warming on the Greenland ice sheet and the West Antarctic ice sheet will likely raise sea level faster than current models now can predict. Remotely sensed and direct measurements of accumulation across the East Antarctic ice sheet showed that current climate models have overestimated accumulation due to snowfall. Cutting-edge radar measurements of the bottom of the East Antarctic ice sheet yield insight on ice sheet origins.

From the polar regions looking into space, IPY allowed for some of the most comprehensive synoptic measurements of the geospace environment ever taken, including new nets for observing and understanding the impacts of space weather on global communications.

Engagement with the inhabitants of the Arctic has led to new capacities for learning about the social processes and health of the people who live in the polar regions.

# 4

# Scientific Tools and Infrastructure

The IPY Vision Report (NRC, 2004) highlighted the view that the polar environment is a tightly coupled system and that "IPY 2007-2008 is an opportunity to deepen our understanding of the physical, biological, and chemical processes in the polar regions and their global linkages and impacts." This system perspective embraced by the IPY research community (Overpeck, 2005) differed from the more disciplinary focus of previous IPY/IGY programs (Wilson, 1961). It also required innovations, such as research tools that enabled simultaneous analysis of multiple climate system components; coordinated use of in situ and remote sensing observational instrumentation; access to nontraditional data sources; and a diversity of new and existing computer models. The approach was inherently multidisciplinary and both facilitated and benefited from international collaboration for the establishment, operation, and maintenance of observing networks and other tools and resources.

The following sections present examples of research tools and their use during IPY, with descriptions of new tools and observatories that were made possible by international organizational arrangements. These developments facilitate the collection of data sets that support examination from a system perspective.

## EXISTING OBSERVATIONS AND PLATFORMS

The large areal extent of the two polar regions makes observing these environments a significant challenge. The areas covered by the Arctic Basin (~12.2 $\times 10^6$ $km^2$), Arctic sea ice at maximum extent (~15 $\times 10^6$ $km^2$), Southern Ocean south of the Polar Front (~35 $\times 10^6$ $km^2$), and Antarctic continent (13.9 $\times 10^6$ $km^2$) are each appreciably larger than the contiguous United States of America (~7.8 $\times 10^6$ $km^2$).

When IPY began, observing stations in these regions had limited space and time coverage. For example, although the longest standard weather record in the U.S. Arctic (in Barrow) dates from 1901, there are vast areas such as the main Arctic Basin and the interior of East Antarctica for which traditional meteorological observations are infrequent or lacking. Therefore, one of the main thrusts of IPY was to establish the capacity to acquire bipolar observational data that could serve as a reference point in long-term examinations of temporal and spatial changes, both in comparison to the earlier IPY and IGY observations and perhaps, more importantly, to those of the future.

### Satellite Observations

Satellite-based observations are an essential component of any program attempting to provide system observations over such large regions. During IPY, data collected from a constellation of satellites (Figure 4.1) contributed to research related to melting or growth of polar ice sheets and of sea ice, elevation changes in sea level, ice-climate feedback loops, cloud heights, aerosol distributions, Earth's radiation balance and temperature, vegetation canopy heights, and global biomass estimates.

FIGURE 4.1 Overview of National Aeronautics and Space Administration (NASA) Earth Observing satellites in operation during IPY. Many sensors carried on these platforms were supported either in full, or in part, by international partners. SOURCE: NASA.

Satellite development and launch is a lengthy process, typically extending over a period of 5 to over 10 years. Thus, by the time the planning of IPY was completed, it was too late to develop and launch satellite systems with a specific IPY focus. Optimization of deployment of satellite-based remote sensing systems related to any broad-based observational program requires that the general program outline and requirements be established well in advance of the actual initiation of the program and be transmitted to the agencies involved in satellite operations. For example, discussions that ultimately led to the ICESat satellite dated back as far as 1979 (Science and Applications Working Group, 1979) (launch date was January 12, 2003). On the other hand, the occurrence and urgency of the IPY strengthened the voice of the polar research community in achieving a rapid recovery from the failed launch of CryoSat-1 in 2005 with the successful launch of CryoSat-2 in 2009.

Satellite observations of the polar regions face several challenges not encountered in lower latitudes. Foremost among these challenges is the lack of data centered on the poles, which arises because Earth observing satellites rarely pass directly over the poles. Rotation of the satellite can avoid this data gap, but this is rarely done; one exception was the Radarsat-1 satellite, which was rotated once on-orbit to obtain synthetic aperture radar (SAR) images of the entire Antarctic ice sheet. These data provide long-term time series of ice sheet changes.

The long periods of darkness and regions of extensive cloud cover, which preclude observations

FIGURE 4.2 Sample of the "you are there" true-color representation of Antarctica possible with the Landsat Image Mosaic of Antarctica (LIMA) draped on a digital elevation model. The region is near McMurdo Station with the Dry Valleys on the right and Koettlitz Glacier on the left. SOURCE: NASA.

that depend on reflected visible light, such as ocean color, provide additional challenges for high-latitude satellite observations. Frequent and consistent satellite observations in these regions require passive or active microwave systems, which are independent of light and cloud cover. However, current passive satellite observations have coarse spatial resolution (tens of kilometers). Active systems such as SAR and scatterometry provide high-resolution all-weather data and have proven to be excellent tools in studies of ice motion and ice type. SAR observations only date from 1991 (ERS-1) and scatterometric studies of sea ice are even more recent.

The development and operation of satellite programs is expensive, making it doubtful that any one country will be able to afford or even have the technical capabilities necessary for launching the suites of satellites required by increasingly sophisticated observing programs. The number of new U.S. Earth observing satellite missions was reduced in the first decade of this century, which set the stage for less extensive observational capacities just as the pace of change increased. As a result, international coordination is all the more important and will undoubtedly be required to focus existing national or foreign satellite capabilities on specific future IPY-like field projects.

A particularly successful execution of this type of international cooperation by multiple national space agencies was the coordination of satellite observing sensors managed by the IPY Space Task Group (STG). The conceptual birth of this project was a Polar Snapshot from Space, which became the IPY project GIIPSY (Global Inter-agency IPY Polar Snapshot Year). It figured heavily in garnering enthusiasm for IPY throughout the scientific communities and funding agencies by demonstrating the clear benefit of collecting consistent data of both polar regions in a short period of time as a unique observational benchmark. A data collection plan, crafted by GIIPSY from multiple scientist requests into a specific set of data requests, was successfully brokered by the STG with 14 space agencies.[1] Once IPY was under way, GIIPSY-orchestrated data were directly responsible for providing a wide variety of compelling IPY projects with essential data, including pole-to-coast multifrequency measurements of ice-sheet surface velocity, repeat fine-resolution mapping of the entire Southern Ocean sea ice cover, a complete visible and thermal infrared snapshot of circumpolar permafrost, snapshots of lake and river freeze-up and breakup, The impressive success of this effort has resulted in the continuation of the STG and serves as much as an important legacy of IPY as the vast wealth of satellite data that were collected during IPY.

The Landsat Mosaic of Antarctica (LIMA[2]), an online atlas and digital library of the continent, is an excellent example of the powerful effect of using large, comprehensive satellite data sets in innovative ways

[1] Participating International Space Agencies: Agenzia Spaziale Italiana, Canadian Space Agency, China Meteorological Administration, Centre National d'Etudes Spatiales (France), Deutsches Zentrum für Luft- und Raumfahrt, (Germany), European Space Agency, European Organisation for the Exploitation of Meteorological Satellites, Instituto Nacional de Pesquisas Espaciais (Portugal), Japan Aerospace Exploration Agency, NASA, NOAA, Russian Federal Service for Hydrometorology and Environmental Monitoring, WMO, World Climate Research Programme-Climate and Cryopshere.

[2] *http://lima.usgs.gov/.*

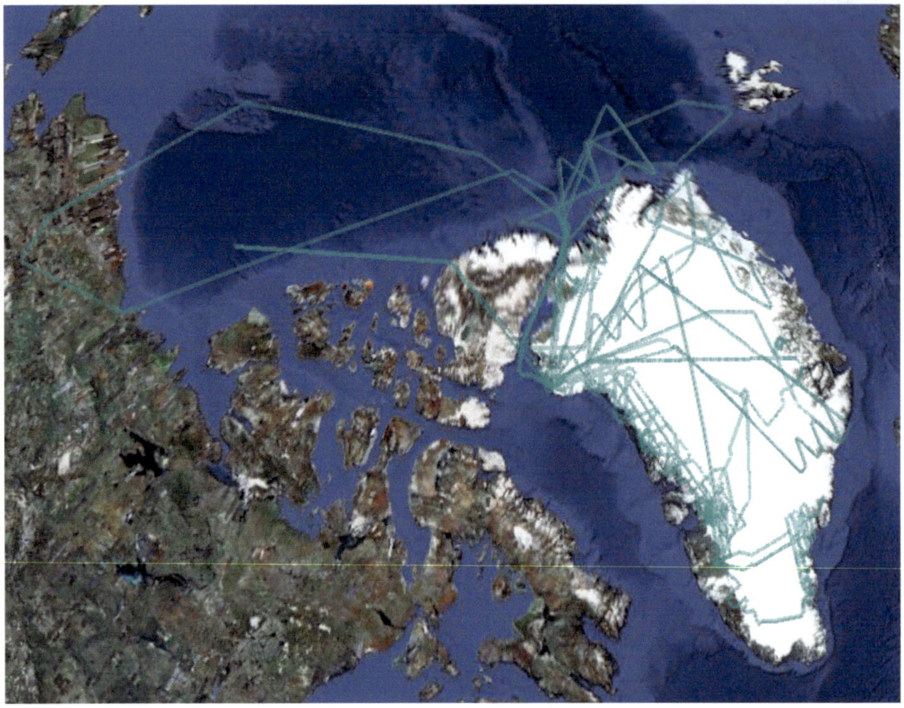

FIGURE 4.3 Tracks of IceBridge flights during March to May 2009 on which data were collected by the Airborne Topographic Mapper, a laser altimetric instrument. Similar patterns of spatially concentrated flights have been completed annually in both the Arctic and Antarctic in subsequent years. SOURCE: NASA.

during IPY. Documenting changes in the distribution of Emperor penguin rookeries was already mentioned as a significant discovery enabled by LIMA, but the manner in which the mosaic was created set a new standard in the production of such continental-scale mosaics (Figure 4.2). This mosaic was created using over 1,000 ETM+ (Enhanced Thematic Mapper Plus) images at 15-meter spatial resolution, which are produced by an eight-band multispectral scanning radiometer on the Landsat 7 satellite. Previous mosaics at this scale served only as an index to the original data and relied on users to return to single images to perform quantitative analysis. LIMA data are scientifically accurate surface reflectances and thus can be used immediately for scientific inquiry, empowering many more scientific users.

In the face of diminished future observational capability, the IceBridge program represents a novel approach to addressing a gap in satellite observational capability at a critical time and is an important addition to the normal NASA program for satellite-based observations of the polar regions. This 6-year aircraft-borne remote sensing program provides an extensive survey of sea and glacial ice masses in both polar regions and is done in cooperation with Australian, British, Canadian, and French investigators. The data collected from this program provide a focused view of a region (Figure 4.3), as opposed to the broader view obtained from satellites. Although the primary purpose of the program is to fill in the data collection gap between ICESat, which provided limited data because of power supply problems, and the launch of ICESat-2 estimated for 2016, it contributed directly to IPY and to the longevity of data collection initiated under IPY.

## Observing Networks

A number of observing networks were established during IPY that attempted to add substantial value to multiple scientific endeavors by combining or extending the data collection capabilities beyond what single countries or projects could either install or sustain. A brief description of some of the different observing networks developed for and utilized as part of IPY follows.

### *Arctic Observing Networks*

In the Arctic, the Sustaining Arctic Observing Networks (SAON) aggregated smaller national observational networks into a broad international coalition. This concept was fostered in the United States as the Arctic Observing Network (AON; a major U.S. government agency IPY initiative, with a majority of the support from the National Science Foundation [NSF]).

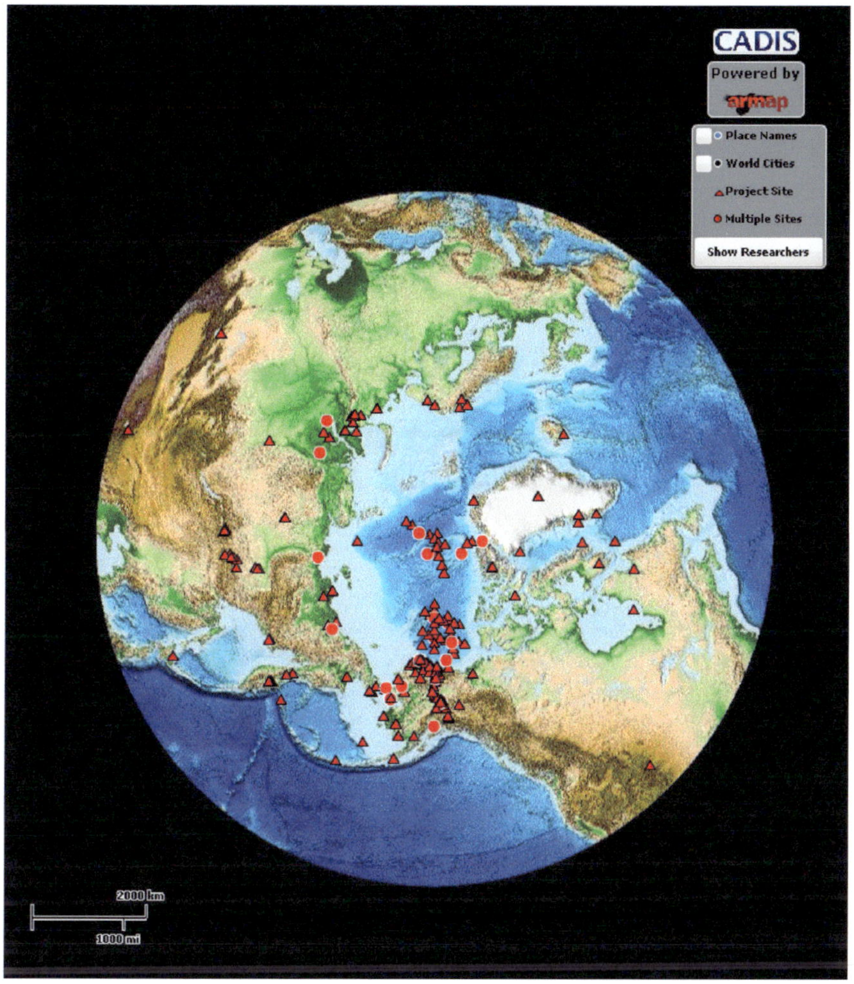


FIGURE 4.4 AON, a major initiative during IPY, encompassed physical, biological, and human observations of the land, ocean, and atmosphere. The geographical coverage of U.S. AON projects can be seen in this map. SOURCE: Arctic Research Mapping Application.

Initially designed to support the data needs of the Study of Environmental Arctic Change (SEARCH) program, AON's multidisciplinary constitution encompassed physical, biological, and human observations (including local/indigenous knowledge) of the land, ocean, and atmosphere. Data products, dissemination, and archiving of AON-collected data are being handled by the Cooperative Arctic Data and Information Service (CADIS[3]) (see Data Management section later in this chapter). The geographical coverage of U.S. AON projects can be seen in Figure 4.4. The international basis of IPY blended the ongoing observations from the U.S.-only AON program with the active international projects under the SAON program. These international collaborations shared many of the same objectives, but with expanded pan-Arctic coverage. The coordination of SAON is still evolving, and researchers are just beginning to publish results[4] with more expected in the coming years.

Separate disciplines within polar science also established broad networks. One example is the integrated Arctic Ocean Observing System (iAOOS), which was formulated by the Ocean Sciences Board and the Climate and Cyrosphere (CliC) Project to explore answers to a number of specific questions concerning the Arctic Ocean and its peripheral seas. The iAOOS framework was designed to collect ocean observations from the seabed to the surface and across both major inflow/outflow pathways between Arctic and sub-Arctic waters as well as a pan-Arctic Ocean transect. As of 2007, over 156 moorings were in place, and these increased to 173 by 2008 (Dickson and Fahrbach, 2011). The regions of interest included the main Arctic Ocean basin; its

[3] *http://www.aoncadis.org/about/aon-projects.htm.*

[4] In particular, several results in Chapter 3 on Marine Ecosystems in a Warming World rely on SAON data.

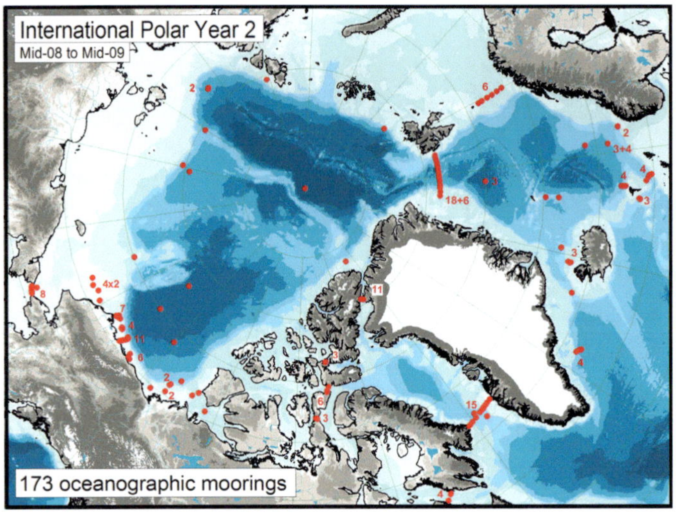


FIGURE 4.5 The integrated Arctic Ocean Observing System (iAOOS) was designed to collect ocean observations from the seabed to the surface and across both major inflow/outflow pathways between Arctic and sub-Arctic waters as well as a pan-Arctic Ocean transect. This map shows the distribution of all 173 current meter moorings and arrays across the iAOOS domain in 2008. Small numerals in red refer to the number of moorings in an array, where these are too numerous to distinguish individually. SOURCE: Dickson and Fahrbach, 2011.

peripheral seas; the Canadian Archipelago; the Fram, Davis, and Bering Straits; Greenland; and the Russian Arctic. The pattern of moorings deployed during IPY concentrated arrays at four major inflow/outflow regions (Figure 4.5). In addition, some of the data sets were collected by remotely controlled SeaGliders (discussed later in this chapter), which is a new technology that will undoubtedly be used extensively in the future. Extensive advance coordination was needed to successfully collect a spatially extensive data set owing to the fact that these regions included all nations bordering on the Arctic Ocean.

Another aspect of iAOOS was the use of autonomous ice-based observatories. The ice-based observatory concept has developed over the last 30 years, but it was significantly advanced during IPY. These assemblages of colocated instruments returned continuous information about the atmospheric boundary layer and surface radiation budget, the evolving snow and sea ice thickness, temperature and salinity profiles, and temporal stress history and deformation, as well as the upper ocean stratification, water properties (including biologically relevant fields), lateral velocity, and mixed-layer turbulence intensity and associated vertical fluxes (Figure 4.6). Large amounts of data were collected via the ice-based observatories and are still being collected.

Without doubt, new thinking has emerged about the role of the arctic seas in climate. For example, the temperature and salinity of waters flowing into the Norwegian Sea at the Scottish shelf and elsewhere along the Kola Peninsula in the Kara Sea are the warmest observed in over 100 years (Dmitrenko et al., 2008; Holliday et al., 2007; Polyakov et al., 2007). New modeling of the flow by Karcher et al. (2007) show that the this warm Atlantic water layer may have less impact on sea ice as it circulates around the Arctic Basin. Instead it may have more influence on the density contrast and height of this water mass where it exists over the Denmark Strait overflow, thus influencing the thermohaline conveyor into the future. Continued monitoring of this evolving system speaks to the legacy this observing system will have to improving our understanding of the sensitivity of the northern seas to climate change.

The overall goal was to deploy observing sites across the Arctic during IPY with the intent that a subset of the sites would remain in operation after IPY. The SAON process built upon the scientific community's experience in the Arctic and clearly facilitated deployment of the various IPY Arctic observing programs. By improving international coordination and planning, SAON helped to provide better coverage and perhaps even improved cost-effectiveness. One important goal was to aid in developing sustained support for such networks beyond IPY because long-term support for monitoring programs has invariably proven to be difficult. It is hoped that the SAON program will contribute to the temporal extension of this data set, thereby ensuring an important contribution to climate change research.

An important new data product, the Sea Ice for Walrus Outlook (SIWO), emerged from these observing networks, and has had a direct and immediate

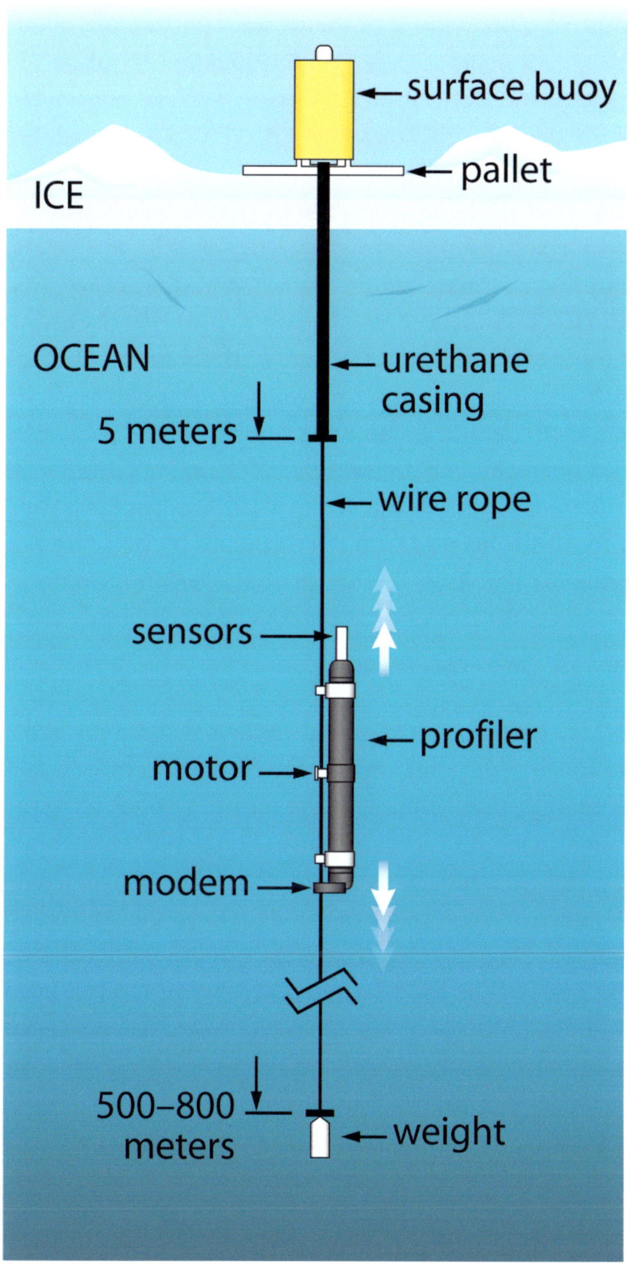


FIGURE 4.6 The concept of the autonomous ice-based observatory has developed over the last 30 years, but it was significantly advanced during IPY. Shown here is a schematic for an ice-tethered profiler (ITP). The ITP system consists of a small surface capsule that sits atop an ice flow and supports a plastic-jacketed wire rope tether that extends through the ice and down into the ocean, ending with a weight (intended to keep the wire vertical). A cylindrical underwater instrument (in shape and size much like an Argo float) mounts on this tether and cycles vertically along it, carrying oceanographic sensors through the water column. Water property data are telemetered from the ITP to shore in near-real time. SOURCE: Richard Krishfield, Woods Hole Oceanographic Institution.

impact on Arctic residents. As a key supporter of the U.S. AON, SEARCH[5] (Study of Environmental Arctic Change), an interagency activity to understand the broad, interrelated changes occurring in the Arctic, brought together a diverse group of specialists to share and discuss sea ice distribution predictions for the anticipated sea ice minimum and facilitate the exchange of approaches and the successes (or failures) of the resulting predictions so that forecasting skill could be improved. The forecasts were then provided to native hunting groups in the form of the SIWO. Ultimately, observations made during hunts were provided to the forecasting group, with the expectation that this would contribute to improved 10-day weather and ice forecasts provided to communities located along the coasts of the Beaufort, Chukchi, and Bering Seas.

### *Antarctic Observing Networks*

In the Southern Hemisphere, planning for a network of ocean observations began during IPY. This project, the Southern Ocean Observing System (SOOS), is designed to provide multidisciplinary observations from one of the least well-observed parts of the ocean. The region is remote, and its climate and associated sea state are sufficiently formidable to limit observations. The goal of SOOS is to utilize the unprecedented level of cooperation available under IPY to gain a synoptic snapshot of the state of the Southern Ocean, using both established tools and new technology such as cryospheric satellites, autonomous profiling floats, and miniaturized sensors deployable on marine mammals. When it is implemented, SOOS is designed to provide the long-term measurements required to improve understanding of climate change and variability, biogeochemical cycles, and the coupling between climate and marine ecosystems. The full SOOS plan describes the combination of sustained observations needed to address the key science challenges identified for the Southern Ocean (Rintoul et al., 2011). SOOS consists of a series of elements that allow these science challenges to be addressed.

The primary SOOS elements are repeat hydrography, enhanced Southern Ocean Argo system, underway sampling from ships, time-series stations and monitoring of key passages, animal-borne sensors, sea ice

[5] *www.arcus.org/search/index.php.*

observations and enhanced ice drifter arrays, observation of ocean circulation under ice and ice shelves, enhanced meteorological observations, remote sensing, and inclusion of observations of lower trophic level processes and ecological monitoring. The general location of proposed repeat hydrography and Argo float observations (Figure 4.7) illustrate the region covered by the SOOS. The general decrease in coverage with increasing latitude is associated with increasing remoteness as well as with the presence of sea ice. An important aspect of the SOOS is emphasis on data archaeology, management, archiving, and access, as described in the SOOS science plan (Rintoul et al., 2011). This is considered critical to understanding any new measurements.

As initially planned, SOOS intends to make use of a full suite of remote sensing observations including radar and laser altimetry, scatterometry, SAR, ocean color, and passive microwave observations. Observations from ICESat were sufficient to verify the use of laser altimetry in obtaining estimates of both snow and sea ice thickness in the Southern Ocean; information that has been largely lacking to date.

Each element of the SOOS observing tools exists in some form and is ready for implementation now. The international effort in coordinating research programs in the Southern Ocean during IPY[6] demonstrated the readiness and feasibility of a comprehensive, integrated system of Southern Ocean observations. In 10 years, the SOOS will likely have expanded to rely more heavily on autonomous sampling that includes floats, gliders, satellites, moorings, and animal-borne sensors. A complete discussion and overview of the SOOS strategy is given in Rintoul et al. (2011).

### *Bipolar Observing Systems*

The Polar Earth Observing Network (POLENET[7]) is a terrestrial, bipolar example of a new observational network established during IPY. Prior to IPY there were very few continuous GPS receivers and seismic sensors in either the Antarctic or Greenland. The establishment of POLENET provided a polar observing system for geodesy and seismology in both regions (Figure 4.8) at a greater density than found at some other remote regions of the world. For example, about 50 autonomous seismic stations operated continuously in the Antarctic interior during IPY as contrasted with one (South Pole) prior to IPY. The continued operation of many of these stations is an important legacy of IPY. Data from these stations will contribute important insights into glacial rebound in both areas, internal processes within the ice, and rates of ice loss. POLENET data in combination with data from the Gravity Recovery and Climate Experiment (GRACE) satellite is leading to improved ice mass balance estimates to determine how the world's largest ice sheets in Greenland and Antarctica are changing (Bromwich and Nicolas, 2010; Wu et al., 2010). Longer GPS time series will also lead to better estimates of the history of ice mass fluctuations since the last glacial maximum.

### *Sea Ice Observing Networks*

Extensive observational networks carry an additional benefit that unforeseen shortfalls in one can sometimes be minimized by shifting some observational tasks to another network. This was demonstrated by the interactions between the U.S. Sea Ice Mass Balance in the Antarctica (SIMBA) and the Australian Sea Ice Physics and Ecosystem eXperiment (SIPEX) programs. The SIMBA program was an international interdisciplinary study focused on sea ice processes integrated with oceanographic, meteorological, and marine mammal components. During 2007, while operating in the Amundsen Sea, a ship fire resulted in the loss of 3 weeks of a planned 2-month field season. However, this loss of data was avoided by transferring some aspects of the SIMBA program to the SIPEX program operating on the other side of the continent. It also was possible to coordinate these changes with the NASA ICESat laser altimeter mission. The resulting satellite-based data set, combined with concurrent field observations, has helped refine algorithms for converting sea ice elevations into more accurate sea ice thickness. These results will contribute to the first analysis of Antarctic sea ice thickness distributions and change. Without previous coordination through IPY, such last-minute changes would have been more difficult to implement.

The experiences in setting up many of the IPY observing networks provide lessons for planners of

[6] See the ICED-IPY website at *http://www.iced.ac.uk/science/ipy.htm*.

[7] *http://www.polenet.org/*.

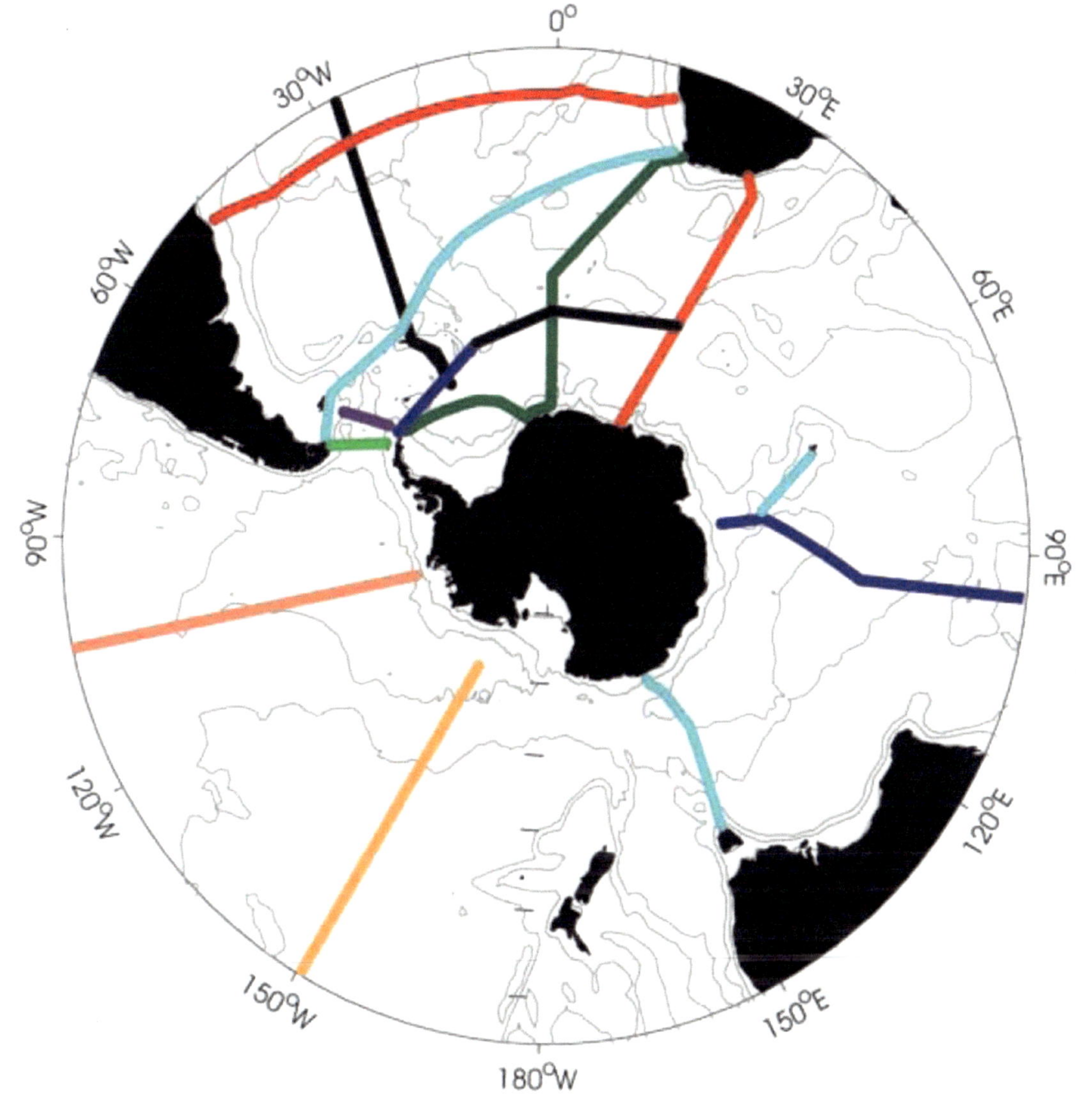


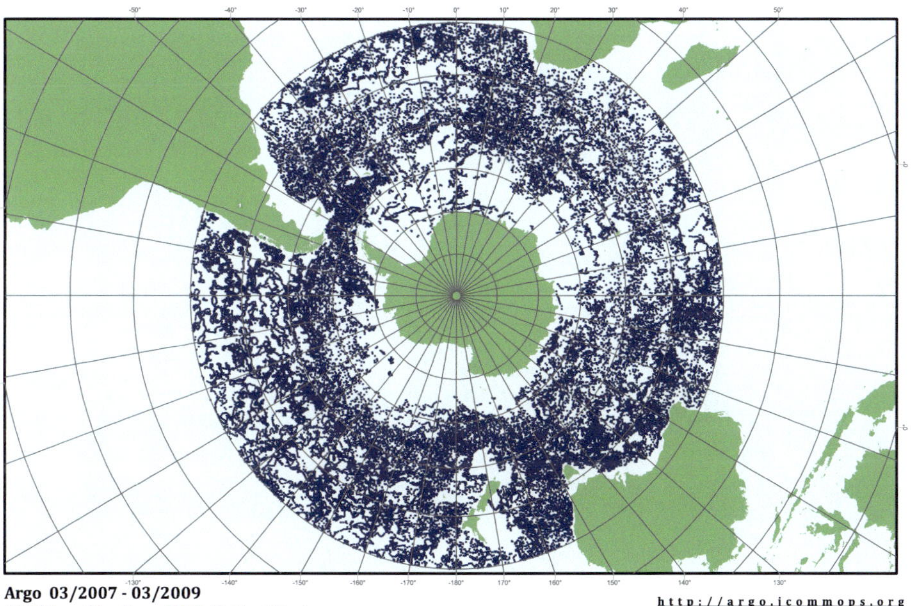


FIGURE 4.7 The Southern Ocean Observing System (SOOS) is designed to provide multidisciplinary observations from one of the least well-observed parts of the ocean. Two key components of SOOS are shown here: repeat hydrographic lines and Argo floats. Top panel: The locations of the proposed SOOS hydrographic lines. Bottom panel: The locations of Southern Ocean Argo float observations made during IPY; the general decrease in coverage closer to the Antarctic continent is associated with increasing remoteness as well as with the presence of sea ice. SOURCE: Rintoul et al., 2011.

related future programs. For example, sensor deployment was a challenge. This will undoubtedly continue to be the case in the future even though environmental conditions may be quite different. If current climate trends continue, areas of multiyear ice, favored as deployment sites, will be both smaller and increasingly remote. Furthermore, although the increase in the ice-free areas during the late summer will favor the use of ships to deploy buoys, subsequent ice formation during the winter may severely limit the usefulness of such buoys. Also, providing a data set that was adequate for all possible users was particularly difficult, for example, the biogeochemical aspects of polar observational programs could and should have been expanded. Thus, it is important to formulate measurement needs well in advance of the actual program so that the collection systems can be optimized for the specific research needs. Involvement of early career scientists in programs such as iAOOS that are expected to last for decades is critical—it may advance the careers of the younger scientists, but it will certainly contribute to the desired longevity of the observing program, which may well exceed the temporal involvement of the initial developers.

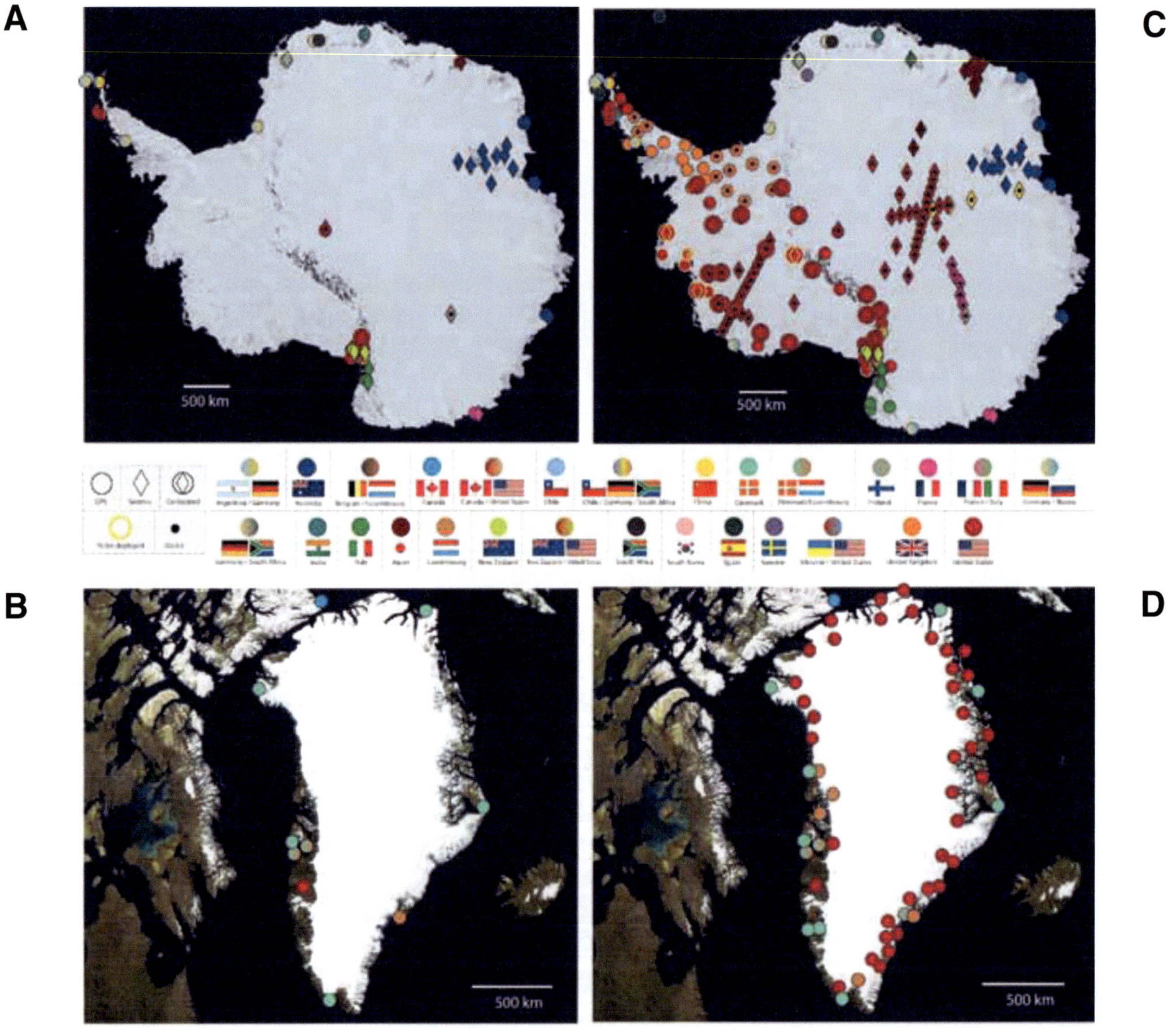


FIGURE 4.8 Maps showing POLENET networks before and after IPY. Continuously recording GPS and seismic stations are shown prior to IPY (a, b) and deployed during the extended IPY period from 2006 to early 2010 (c, d). SOURCE: POLENET database; maps drafted by M. Berg and S. Konfal.

### Icebreaker Support Capabilities

Performing scientific research in the polar regions often requires research vessels capable of operating in fully and partially ice-covered seas. As such, ships capable of breaking through ice are required for operations in both the Arctic and Antarctic. Icebreakers are required for scientific and supply missions in the Arctic, including operations in the Bering, Chukchi, and Beaufort Seas. In the Antarctic, heavy ice-breaking capability is required to clear a path through the thick continuous first-year (FY) sea ice in McMurdo Sound to allow cargo ships to carry out the annual resupply of the U.S. and New Zealand bases at McMurdo. This resupply is essential to the operation of these stations and the U.S. station located at the South Pole.

The current status of U.S. ships with icebreaking capabilities is uncertain. The USCG icebreaker *Polar Sea* is scheduled to be decommissioned in 2011. Its sister ship, the *Polar Star*, is now over 30 years old and has been undergoing repairs designed to extend its service life for 5 to 7 years. The two ships are unique in that they have the capability of increasing their horsepower from 16,000 to 60,000 by activating an additional gas turbine system, which allows them to complete the McMurdo break-in.

The two other U.S.-owned ice-capable vessels in current use are the USCG *Healy* and the NSF chartered RVIB *Nathaniel B. Palmer*. The *Healy*, although the largest ship in the history of the Coast Guard, has proven to have great difficulty maneuvering in thick continuous FY ice; therefore its operations are currently limited to the Arctic where it has proven to be an effective asset for operations in the Bering, Chukchi, and Beaufort Seas. The *Palmer*, operated on contract for the NSF, was designed for Antarctic operations and has proven to be capable of operations in FY Antarctic pack ice, but its ice-breaking capabilities are less than required for the annual break-in for the McMurdo Sound. The *Palmer* entered service in 1992 and has been in essentially continuous operation for 19 years. As a result, it will undoubtedly require enhanced maintenance during its remaining operational lifetime.

During IPY, NSF was able to make arrangements to have the Swedish icebreaker *Oden* break out the channel to McMurdo Sound. In the past, Russian icebreakers have also been chartered to carry out this essential task and it has just been announced that during the coming 2011-2012 austral summer another Russian icebreaker, the *Vladimir Ignatyuk*, has been contracted for this task. Such arrangements, if possible, are desirable in that they, by leveraging other country's marine investments, provide both nations with a return that is greater than would be obtained by working alone. The difficulty is that it is impossible to guarantee that such arrangements can be made every year. If a suitable replacement cannot be found, U.S. and New Zealand operations in the McMurdo area and in East Antarctica in general will have to be severely curtailed, if not completely cancelled.

In the Arctic, the *Healy* will continue to operate, and the *Sikuliaq*, with less icebreaking capabilities than those of the *Healy*, is under construction. At 261 feet and 5,750 HP, the *Sikuliaq's* winter in-ice operations will probably be limited to the Bering Sea, and its summer operational area will likely include both the Chukchi and Beaufort Seas. It should prove to be an effective operational platform but it is not designed for deep extended operations into the Arctic Basin. The *Healy* may prove capable of winter operations in the main Arctic Basin, but confidence in this capability awaits further field tests.

In the Antarctic, as the *Palmer* approaches the end of its operational life, the U.S. Antarctic Research Program will essentially be without icebreaking capability. In particular, it will lack the capability of operating in the Southern Ocean during the winter and of breaking out the U.S. Base at McMurdo Sound during the summer, an operation essential to base resupply. If U.S. operations are to continue in the Antarctic, the lengthy design and construction time required to produce an operational icebreaker requires that the decision to initiate development of a replacement ship should be made soon. Inherent in the design should be two primary capabilities. This replacement vessel needs to be capable of operating in thick, continuous FY ice and should have a design that supports multidisciplinary research.

## NEW OBSERVATIONS FROM SPECIFIC TOOLS

Several recently developed research tools were used during IPY. Given the long lead time required to plan, secure funding, and develop new tools, work on all of these research tools began well before IPY and in some cases even before IPY planning began. While IPY may not have directly benefited the initial stages of tool development, the existence of a large, multi-disciplinary, international project such as IPY provided the opportunity to deploy and test new tools as parts of large observing networks. Some examples of such innovative tools that provided a new perspective on the polar system are highlighted below.

### SeaGliders

Over the past 25 years, much has been learned about the spatial and temporal variability of the physical, chemical, and biological distributions and processes in the ocean, and it is now known that changes in oceanic processes occur over a variety of space and time scales. In high-latitude systems, seasonal cycles in irradiance, wind fields, and sea ice concentration and extent are important in determining ocean stratification, upper ocean heating, and biological productivity. However, short-term physical variations such as event-scale mixing also play a dominant role in controlling the magnitude of these processes as well as their timing and duration. These events are likely quite common in high-latitude systems, but are rarely sampled, and as such their importance to seasonal and annual cycles remains poorly quantified (see the section on Sea Ice in Chapter 3).

One reason such events are rarely sampled is that traditional ship-based sampling does not often resolve processes on either the mesoscale (i.e., eddies, fronts, and currents), or seasonal time scales. To adequately resolve these sampling needs, new technologies were needed. Gliders, autonomous vehicles that can sample a limited suite of variables for long periods of time, represented one such new technology. Gliders are capable of extended missions (several months), are durable, and carry sensors that provide coincident measurements of temperature, salinity, optics, fluorescence, and transmissometry, which provide high-resolution space and time representations of environmental and biological conditions.

During IPY, SeaGliders (Figure 4.9), developed by U.S. scientists, were deployed off western Greenland. These gliders provided 6-month time series of freshwater export from the Arctic Ocean through the Canadian Arctic Archipelago and Davis Strait into the Labrador Sea. The amount of freshwater entering the Labrador Sea determines the formation of dense water in this region. Thus, understanding the variability in freshwater supply is critical to understanding the effect of climate change on the larger-scale ocean thermohaline circulation. The SeaGliders were one of the many instruments included in the Arctic Observing Network (AON). The SeaGlider represented an advance over other autonomous underwater vehicles in that it does not use propulsion to move through the water, can operate on its own for several months, and can operate effectively in ice-covered regions. Unlike ARGO floats, it can maneuver in vertical and horizontal space. At specified intervals it sends data and its GPS position via the iridium satellite to a computer base station, allowing almost real-time analysis of the data. The glider can then receive new instructions from the operator at the base station about what to target next (position), how deep it should go, and its data sampling frequency.

IPY showed that SeaGliders are viable tools for measuring remote regions at space and time scales that are relevant to the oceanographic processes that are of concern for climate change. Subsequent to IPY, SeaGliders had been deployed successfully in other high-latitude regions, such as the Ross Sea. This instrumentation is rapidly gaining acceptance and is becoming part of multidisciplinary oceanographic programs.

### Animal-Borne Ocean Sensors

The use of animal-borne conductivity-temperature-depth satellite relay data loggers (CTD-SRDL) tags to study habitat and behavior was the core theme of the IPY MEOP (Marine Mammals as Explorers of the Ocean Pole to Pole) program. This international effort deployed tags in large numbers on grey and hooded seals in the Canadian and Norwegian Arctic and on crabeater, elephant, and Weddell seals in the Southern Ocean (Figure 3.13). The deployment of 85 CTD-SRDLs on southern elephant seals as part of the international Southern Elephant Seals as Oceanographic Sensors (SEAOS) provided a circumpolar assessment of the behavior of southern elephant seals as well as a synoptic view of the oceanography of the Southern Ocean (Biuw et al., 2007).

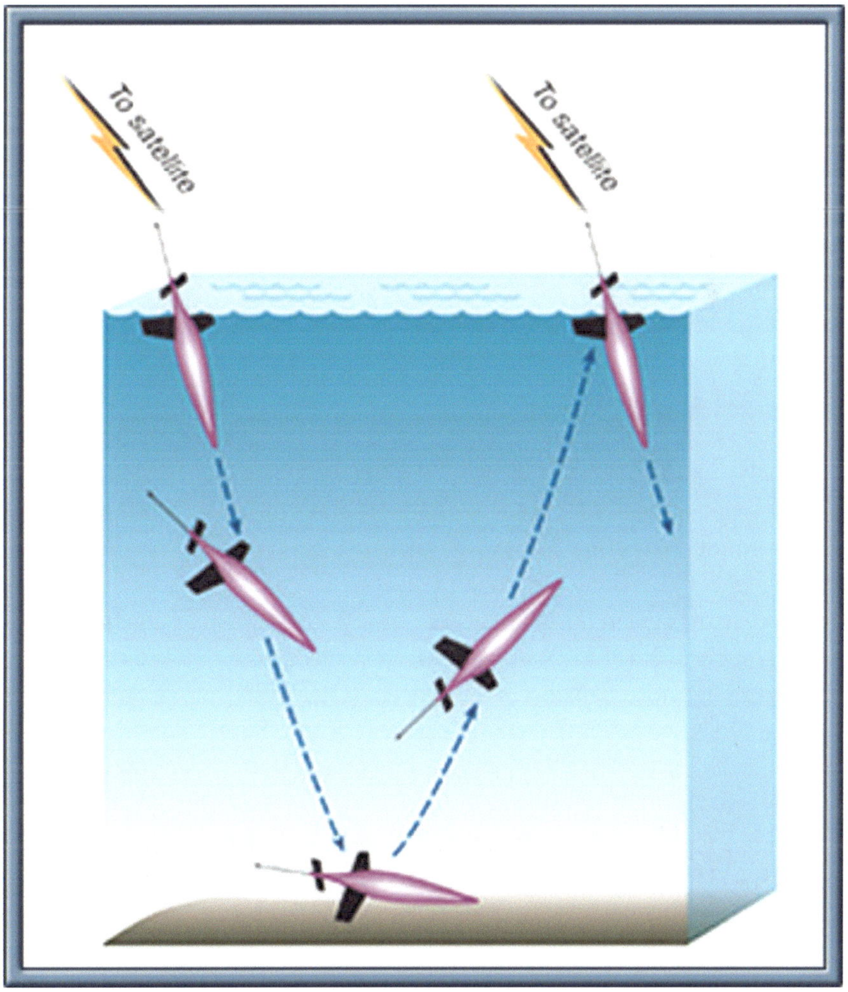


FIGURE 4.9 Photograph of SeaGlider and schematic of SeaGlider operation and communication. SOURCE: Applied Physics Laboratory, University of Washington.

Satellite-linked dive recorders that sample the temperature and/or salinity while simultaneously recording information on the diving patterns of the seals (Biuw et al., 2007) offer a significant advantage over remotely sensed data in that they acquire oceanographic-quality data at a scale and resolution that matches the animals' behavior. An equally important advantage is that these tags provide oceanographic data that can contribute to a better understanding of physical oceanography, especially in areas where traditional shipboard and Argo float coverage is limited or absent, such as the Arctic and Southern Oceans.

In the Southern Ocean, the 2 years of IPY-MEOP sensor deployments resulted in 68,000 CTD profiles collected by 166 seals. These data were from mostly data-sparse regions, and a large proportion was from high-latitude regions in winter in areas with 80 percent or more sea ice coverage. The MEOP program provided comprehensive, synoptic coverage that allowed investigation of factors determining habitat selection and use by important polar marine mammal species. The combination of oceanography and marine mammal ecology in MEOP has significantly advanced understanding of how top predators use their environment as well as providing high-resolution hydrographic measurements that can be used to advance understanding of climate processes (e.g., Costa et al., 2008). The IPY effort provided the large-scale verification of the validity of "animals as oceanographers."

### Unmanned Aerial Systems

Early deployment of Unmanned Aerial Systems (UAS) in polar regions began in the late 1990s with flights conducted from Barrow, Alaska. using Aerosonde UAS. The deployment and use of UAS for polar scientific research expanded during IPY through projects such as the U.S.-Norway Traverse project,[8] the Center for Remote Sensing of Ice Sheets (CReSIS[9]), and the NASA-funded Characterization of Arctic Sea Ice Experiment (CASIE), which took place on Svalbard in July 2009. This project used the NASA Sensor Integrated Environmental Remote Research Aircraft (SIERRA) UAS (Figure 4.10) to observe sea ice roughness and sea ice breakup in high northern latitudes. Similar to SeaGliders, UAS provide a tool that allows for repeat monitoring of atmospheric and surface state in remote or difficult-to-reach locations and provide observations on spatial and temporal scales that allows for investigation of mesoscale features not often observed with other observing systems. Deployment of UAS in the polar regions faces difficulties because of the harsh climatic conditions but avoids logistical difficulties associated with flight clearances needed in more populous mid-latitude locations.

### Automatic Weather Stations

The development and deployment of reliable automatic weather stations (AWS), starting in the early 1980s, helped fill significant gaps in surface weather observations in the polar regions. Extensive AWS networks in the Antarctic (Figure 4.11) and Greenland were in place during IPY. Observations from more than 50 AWS systems are used for operational weather forecasting and thus contribute to many field-based polar research activities by facilitating more accurate forecasts. The data from these networks are readily available[10] and are used for a wide range of research including atmospheric science and glaciology. The Antarctic continent AWS network is maintained through a 12-nation international effort.

### Paleoclimate Tools

Climate change over the last century has been dramatic and well documented (IPCC, 2007b), but is best evaluated in the context of natural climate variability. While changes of the last few decades can be weighed in the short term against instrumental and historical data and observations, a better understanding of the character and pacing of natural Earth system variability can only be assessed through paleoclimate studies. Long continuous geologic records contained in ice cores and sediment cores from both marine and lacustrine depositional systems at the poles provide information about natural variability on a variety of time scales. The data from these cores suggest forcing mechanisms and feedbacks at work in global and regional systems. Modeling linked with the geologic record provides

[8] *http://traverse.npolar.no/.*

[9] *https://cms.cresis.ku.edu/.*

[10] *http://amrc.ssec.wisc.edu/.*

FIGURE 4.10 NASA SIERRA UAS on the runway at Ny Alesund, Svalbard. SOURCE: Julie Brigham-Grette, University of Massachusetts, Amherst.

further information about the sensitivity of regional systems to change caused by human activities.

IPY provided the impetus for a number of international programs and initiatives aimed at recovering exceptionally long, high-resolution geologic records of change in the high latitudes (see Chapter 3 on Discoveries). Because of the exceptional logistical and technical challenges in obtaining these records, new instrumentation and geological drilling systems were designed for the benefit of the polar science community. These systems provide a benchmark for IPY that celebrates collaborations as well as innovation, as illustrated by the following examples.

Drilling successfully at Lake El'gygytgyn (Lake E) in remote northeastern Russia represented a massive logistical and political undertaking (Melles et al., 2011; Figure 4.12). Multiple shipping containers originating at various institutions around the world were transported by ship, rail, truck, and eventually bulldozer. Drilling platform preparation required that the lake ice be artificially thickened to about 2.3 m to support the 100-ton weight of safe operations. Cores were taken using a newly designed hydraulic/rotary system consisting of a diamond coring rig positioned on a mobile platform that was weather-protected by insulated walls and a custom made tent atop a 20-m-high derrick. The system was permanently imported into Russia, where it is now available as a research tool for scientific drilling projects at no cost to the science community until 2014. (Results from this project are described in Chapter 3 in the section on Evidence of Past Climate Change over Geologic Time Scales).

In Antarctica, the new multinational Antarctic Drilling Program (ANDRILL) was launched on the McMurdo Ice Shelf (MIS) in austral summer of 2006 to 2007 by overcoming similar logistic challenges. Using a custom-built drilling system consisting of a diamond drill rig and jack-up platform, the first ANDRILL drilling system operated effectively atop an 85-m-thick portion of the ice shelf that moved laterally and was subject to tides and poorly studied subshelf currents (Figure 4.13). The MIS project successfully recovered 1,285 m of sediment, documenting for the first time the complex interplay among the WAIS, EAIS, and the Southern Ocean (Naish et al., 2009).

Refined technology and geological drilling techniques in both polar regions allowed for the recovery of these unparalleled records, which will catapult the understanding of high-latitude climate evolution over millions of years.

In parallel with geologic drilling, ice core drilling for paleoenvironmental records (Figure 4.14) also made a number of milestones for IPY. The ongoing recovery of a unique 100,000+-year high-temporal resolution record from the interior of West Antarctica (the first attempted there since the Byrd core during IGY) required the development of the DISC (Deep Ice Sheet Coring) Drill by the U.S. Ice Drilling Design and Operations (IDDO) group at the University of Wisconsin, Madison, a drill with unprecedented ability to drill diagonally from selected depths deep in the ice sheet to retrieve additional, replicate ice cores from scientifically interesting depths (Johnson et al., 2007; Mason et al., 2007; Mortensen et al., 2007; Shturmakov et al., 2007). Moreover, the project led to the development of a new field-based system to take multitrack conductivity measurements of the ice cores,[11] which allows scientists to focus on specific core sections, maximizing the scientific information per unit cost. The WAIS Divide project also led to the development of ultramodern laboratory equipment for the continuous stable isotope analysis (oxygen and hydrogen) of recovered ice (Gupta et al., 2009) and carbon dioxide measurements of the "fossil air" enclosed in compressed air bubbles in the ice.

FIGURE 4.11 John Cassano working on an automatic weather station. AWS networks in place during IPY helped fill gaps in surface weather observations. SOURCE: John Cassano, University of Colorado Boulder.

These new technologies were born of longstanding international and science-industry collaborations. Ice coring science had its origins in the IGY era, when the very first ice core was drilled in Greenland by the U.S. Army Cold Regions Research and Engineering Lab. Even though the science and technology are relatively young, international sharing of drill designs and engineering expertise is a hallmark of the ice coring community (Langway, 2008). The International Partnerships for Ice Coring Science,[12] one of the IPY legacy organizations, has a group of international drilling engineers who periodically meet to share knowledge and experience about ice core drilling.

## MODELS AND REANALYSES

Leading up to and during IPY, polar regional modeling began to focus on system modeling. A series of international workshops from 2007 through 2009 culminated in the publication of *A Science Plan for Regional Arctic System Modeling* (Roberts et al., 2010) that highlighted the needs for regional Arctic system modeling (ASM) (Figure 4.15). Another focus of polar modeling during IPY was the development of the Arctic System Reanalysis (ASR) (Bromwich et al., 2010). Regional models for the polar regions, including Polar WRF,[13] have seen increased development in the years since IPY.

Both ASM and ASR activities provided an opportunity for synergies between observational and modeling communities. The polar system focus of these modeling efforts aligns closely with the system focus of IPY. Observations collected as part of IPY provide validation for these modeling efforts and in the case

[11] Kendrick Taylor, Desert Research Institute, personal communication, 2011.

[12] *http://www.pages-igbp.org/ipics/.*

[13] *http://polarmet.osu.edu/PolarMet/pwrf.html.*

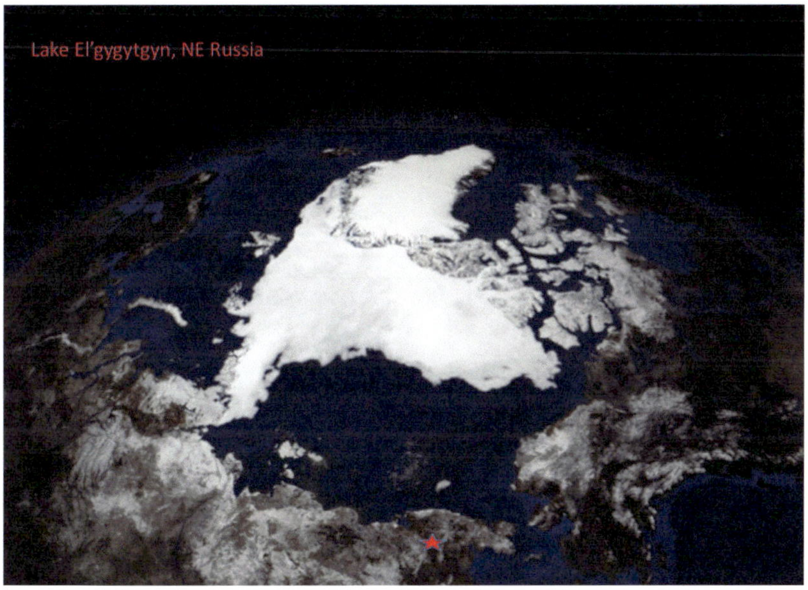


FIGURE 4.12 The U.S.-Russia-German-Austrian Lake El'gygytgyn Scientific Drilling program recovered the first continuous record of past climate change reaching back to 3.6 million years. Images show the field site and the Lake El'gygytgyn Science Party. Lower left map shows the location of the field site. SOURCES: Lake El'gygytgyn Science Party; map: Brigham-Grette et al., 2011.

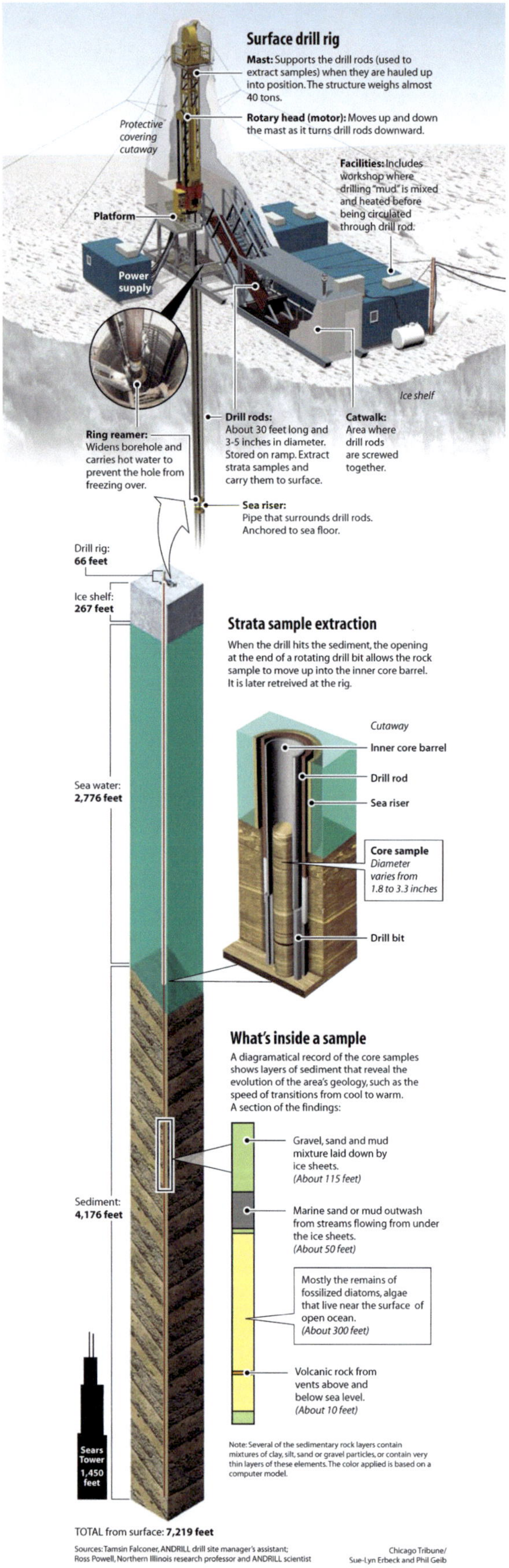


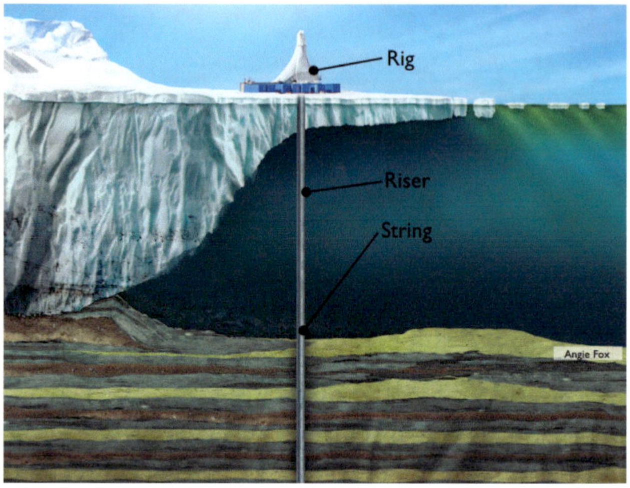


FIGURE 4.13 Schematic of the ANDRILL-MIS drill core. SOURCES: left, Chicago Tribune; right: ANDRILL Science Management Office, *http://www.andrill.org*.

of the ASR provided data to constrain the reanalysis. Results from modeling efforts such as this can highlight regions of interest for future observational efforts and can represent in a physically consistent way processes acting on small spatial and temporal scales which until recently have been poorly sampled.

Overall, IPY facilitated closer integration between the observational and modeling communities. Data assimilative models provided inputs used to guide observation and deployment activities and these in turn provided data for the models. This integration

FIGURE 4.14 A 1-m-long section of ice core from the West Antarctic Ice Sheet Divide Ice Core; section contains a dark ash layer. SOURCE: Heidi Roop.

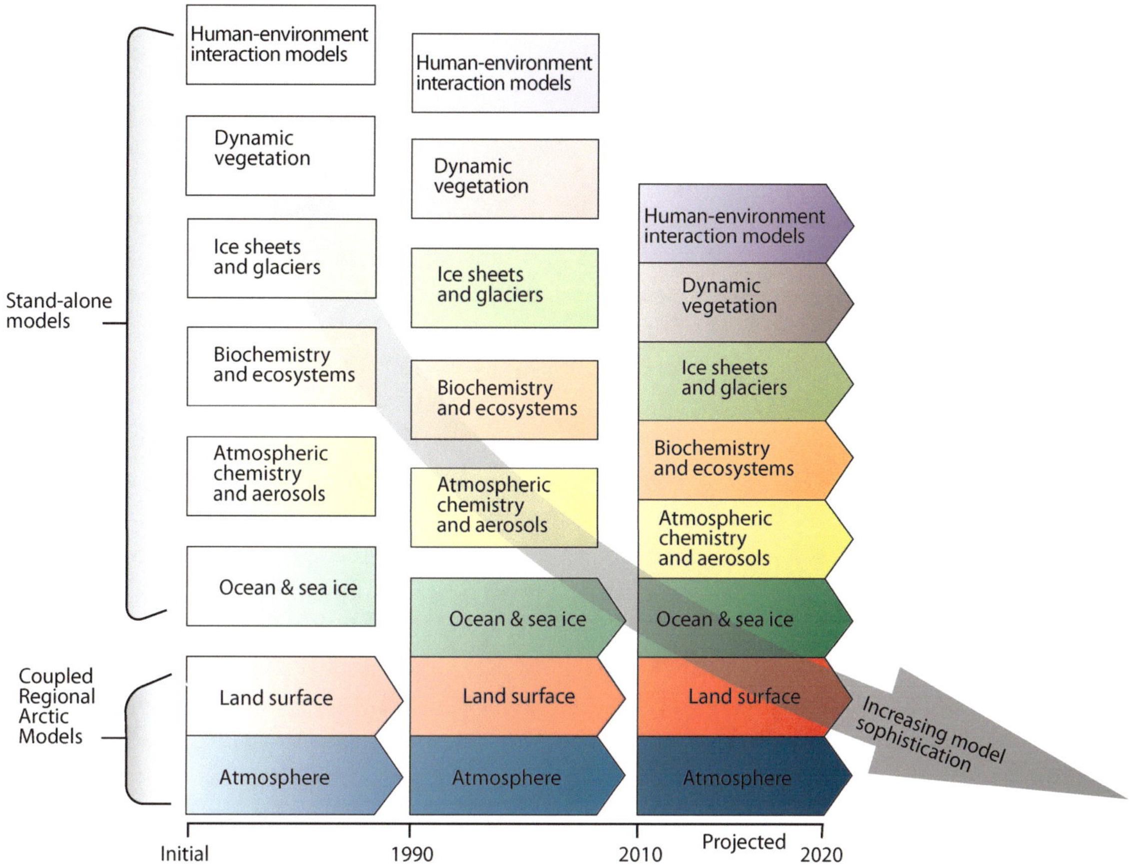


FIGURE 4.15 Evolution of regional arctic models. Geophysical ocean-sea ice-atmosphere-terrestrial components have progressively been coupled during the past two decades, while other components have been studied as stand-alone systems. The proposed Arctic System Modeling program will bring a greater understanding of interconnectivity within the Arctic by fostering coupling of biogeochemistry and human dimensions components. SOURCE: Roberts et al., 2010.

represented a positive interaction and as such provides an important legacy from IPY.

## DATA MANAGEMENT

The IPY framework report underscored the central role of data by stating "In fifty years time the data resulting from IPY 2007-2008 may be seen as the most important single outcome of the programme" (Rapley and Bell, 2004). Rapid changes in the polar regions make the need to share data more acute because the knowledge being urgently sought to inform decisions is well beyond the means of single investigators, projects, or even single countries. While no data management policy was put in place before IPY project proposals were submitted to the Joint Committee, the Joint Committee later stressed the importance of data sharing and management by instituting a data policy for IPY that emphasized the need to make IPY data, including operational data delivered in real time, available "fully, freely, openly, and on the shortest feasible time scale." An emphasis on data availability, sharing, and thus management, could have placed IPY on the forefront of efforts to make fundamental advances in this area. However, strategic differences in Arctic and Antarctic data management and the conduct of science investigation within new interdisciplinary structures all challenged data management during IPY. There were no internationally coordinated IPY planning efforts to engage funding sources for planning data archiving, which created an additional funding problem for post-IPY international archiving. Moreover, during the buildup to IPY, the International Council for Science's (ICSU's) assessment of the world data centers (many of which were established during the IGY) questioned their viability and collaboration, recommending a major overhaul of ICSU data structures (ICSU, 2004). Thus, ICSU viewed IPY as an opportunity to make critically needed advances in data management.

IPY's scientific success depended on handling data in ways that enabled researchers to access and use various data sources and in novel ways. The IPY Joint Committee addressed the needs of improved data management by forming a Data Policy and Management Subcommittee[14] in late 2005 whose task included the generation of an IPY Data Policy and Strategy. This policy and strategy expressed the importance of data sharing and publication, interoperability across systems through the establishments and adherence to data standards, sustainable preservation and stewardship of diverse data, and coordinated governance to ensure access for all researchers. All IPY projects pledged to honor this policy—a remarkably universal expression of recognition of the important role data played in current and future polar research. This subcommittee also defined an IPY metadata profile consistent with what was being used at several polar data centers and the Global Change Master Directory (GCMD). It also requested national IPY organizations to name a data coordinator responsible for promoting the IPY data standards in their respective countries. To date, 16 countries have identified a data coordinator; in the United States this responsibility is with the National Snow and Ice Data Center (NSIDC).

As a result of IPY, several data centers have established pilot projects to exchange metadata records using the IPY profile and the Open Archives Initiative Protocol for Metadata Harvesting (OAIPMH). Metadata from centers in Canada, Norway, Sweden, the United Kingdom, and the United States are directly provided to the GCMD. With support from the NSF, an IPY Data and Information System (IPYDIS) was established in collaboration with the Electronic Geophysical Year (eGY[15]). The eGY was an independent effort focused on making past, present and future geophysical data rapidly, conveniently and openly available. It was supported internationally by the International Association of Geomagnetism and Aeronomy and the International Union of Geodesy and Geophysics, while US-based support was provided by NASA, NSF, the U.S. Geological Survey, and the Laboratory for Atmospheric and Space Physics, University of Colorado, with in-kind contributions from the American Geophysical Union and the National Center for Atmospheric Research.

IPYDIS[16] established a global partnership of data centers, archives, and networks to ensure proper stewardship of IPY and related data. The substantial U.S. funding support of IPYDIS demonstrates the U.S. commitment to sound data management for internationally collaborative science now and into the future. NSIDC coordinated IPYDIS. It was guided by the IPY policy set by the IPY Data Policy and Management Subcommittee and requested updated data-related information from all IPY projects as those projects evolved. The website[17] has instructions to guide researchers submitting metadata to either nationally designated IPY data centers or to the more general GCMD-IPY portal where data are organized into 14 disciplinary categories. IPYDIS also guided users interested in accessing data to specific sites either with direct links or through a prototype search interface called Discovery, Access, and Delivery of Data for IPY (DADDI[18]). DADDI is presently limited to Arctic coastal data. Under the definition of IPY data established by the IPY Data Subcommittee, 1,400 data sets resulting from IPY have been catalogued in the GCMD (Parsons et al., 2011). Also part of IPYDIS, the International Polar Year Publications Database (IPYPD) was created by the Arctic Science and Technology Information System (ASTIS), the Cold Regions Bibliography Project (CRBP), the Scott Polar Research Institute (SPRI) Library, the Discovery and Access of Historic Literature of the IPYs (DAHLI) project, and the National Information Services Corporation (NISC). It is intended to serve as a database for all publications related to IPY.

In addition to the more general repository effort of IPYDIS, many large IPY projects constructed their own data portals, for example, the Antarctic Drilling Project, the Arctic Observing Network, the Circumpolar Biodiversity Monitoring Programme, the Polar Earth Observing Network, and the Scientific Committee on Antarctic Research Marine Biodiversity Information Network. These additional portals provide access to data not yet available through GCMD, but do demonstrate timely release of data. Other data collected

[14] *http://classic.ipy.org/international/joint-committee/data-management.htm.*

[15] *http://egy.org/index.php.*

[16] *http://www.ipydis.org/.*

[17] *http://gcmd.gsfc.nasa.gov/KeywordSearch/Home.do?Portal=ipy&MetadataType=0.*

[18] *http://www.nsidc.org/daddi/.*

during IPY by sensors designed and operated with a mission larger than polar-only were naturally added to existing data repositories already designed specifically for those sensors. An example of this category is the wealth of satellite data of the polar regions orchestrated by the IPY Space Task Group that was established for the purpose of coordinating space agency planning, processing, and archiving of the IPY Earth Observation legacy data set. Other sources of data include repositories for polar materials and samples such as the U.S. National Ice Core Laboratory,[19] the U.S. Polar Rock Repository,[20] the Antarctic Marine Geology Research Facility,[21] and the U.S. National Lacustrine Core Facility.[22] The collections are available for a wide range of scientific research including current and future studies.

The great diversity of IPY data resulted in some barriers to its use, some causes of which included data unfamiliar to scientists outside the originating discipline, data that did not fit into the organizing structure of the data center, or the absence of tools either to work with data or to even locate relevant data. In some cases these barriers were overcome by the investigators themselves; for example, for the IPY project "Antarctic Snow Accumulation and Ice Discharge (ASAID)," customized software code as well as formal documentation describing its use was supplied to NSIDC. Often, the data center is expected to develop and improve tools so that investigators submitting data (or metadata) find it easy to describe and transfer their data and investigators seeking data can efficiently search for relevant data sets. No grand solutions were achieved during IPY, but the net result of IPY's focus on data, data sharing, and data management prompted many constructive steps by data centers and investigators alike that have improved data accessibility. Continued efforts in this arena are essential.

## Lessons and Legacies in Data Management

The IPY Joint Committee report stated that "the IPY policy of general openness built from existing policies appears to be an initial success" (Parsons et al., 2011). The principle that as much data as possible be fully available in the public domain was adhered to by many and continues to enable discovery-driven research. However, the breadth of IPY social science projects highlighted that when human subjects are involved, other considerations related to data description and availability need to be considered, lest the effectiveness or accuracy of the research be compromised. In social science, trust needs to precede data acquisition; the building of relationships with local residents depends on how collected information will be used and distributed. This is not new to social scientists, but less familiar to physical scientists who, through IPY, sought to bridge the gap between social and physical sciences. On the other hand, the integration of physical science research goals into projects that included local residents often provided a demonstrable and tangible benefit to the residents, building trust and reinforcing the notions that the research was synergistic and that data sharing is, in fact, an equitable enterprise.

The step-change increase in understanding polar systems attempted by IPY highlighted the central role that data and its management play in achieving that goal. There have been challenges in sharing and archiving data (Carlson, 2011), and the IPY experience illustrated that data handling is most successful when nations commit program resources to each phase of the data's life (e.g., collection, reduction, distribution, and archiving). Experts in data management are critical members of any team attempting internationally coordinated science on the scale of IPY.

The contrasts between the Arctic and the Antarctic reflect onto the differences in how data are managed, increasing the difficulty of bipolar research. There are geopolitical and social dimensions in the Arctic that complicate data management and accessibility. National interests are stronger, cultural and health issues abound, and each is changing rapidly as the Arctic physical environment changes. The International Arctic Science Committee (IASC) and SCAR are the natural organizations to coordinate data management in the two poles, but presently they do not have a consistent data policy. The IPY data subcommittee suggested a new CODATA Task Group to help plan a transition from IPYDIS to relevant international data structures and organizations and also recommended that IASC and SCAR work with this task group to create a single polar data policy and associated data management procedures and structures.

---

19 *http://nicl.usgs.gov/.*

20 *http://bprc.osu.edu/rr/.*

21 *http://www.arf.fsu.edu/.*

22 *http://lrc.geo.umn.edu/laccore/.*

The effort exerted during IPY toward networking various data centers should continue and expand into the future. Today, the rapid exchange of vast data volumes allows a distributed data center, and IPY used this to advantage in linking data sets hosted in widely separate data centers to form much larger virtual data centers (also the eGY concept). This trend will continue, but the partnerships between data centers need to be more than electronic links. To fully serve the needs of scientists strong in a single discipline but interested in multiple disciplines, the form and format of the data sets need to be modified to enable an increased level of interdisciplinary research. This will require collaborative planning on the part of data managers and scientists. It may not require changing the actual form of the data, but rather provision of interface tools that allow a data set to be understood by a variety of disciplinary experts.

Other considerations include institutional requirements for data release. Parsons et al. (2011) expressed the view that "the experience in IPY... has shown that most effective enforcement mechanism is through funding mechanisms that either withhold some funding or reduce the ability of scientists to obtain future funding opportunities if they do not adhere to the data policy." This is a familiar condition in the United States, where the NSF, which funds much of the polar research, imposes just such a requirement on funded investigators. In return for data shared by investigators and data managed by data centers, it is very important that users of the data provide proper and complete acknowledgment and credit these data in their subsequent use. Guidance for proper citation supplied by data centers is becoming more common. For example, with data sets archived in the National Snow and Ice Data Center, there is a sentence provided that explicitly states how the data and its archive should be referenced in documents that make use of the data. Understanding the data policies of all government funding entities involved is an important component for planning future international science endeavors like IPY.

## CONCLUSIONS

The polar regions have always presented great logistical challenges because the terrain is vast, access can be difficult and expensive, the working conditions are invariably difficult, and the areas of interest frequently cross national boundaries. As a result, to adequately observe the large-scale systems interacting, international cooperation is frequently a necessity, requiring significant planning. Observation networks such as SAON, iAOOS, and SOOS developed and/or expanded during IPY. IPY put in place the planning and infrastructure needed to develop long-term sustained measurement systems for the Arctic and Antarctic. The structure of these networks will continue to evolve as the data are analyzed and needs change. However, sustaining these systems in the long term will continue to present a challenge to the research community.

IPY saw numerous examples of first-time deployments of new tools for observing the polar climate, ecosystems, and beyond; examples include SeaGliders, unmanned aerial systems, and animal-borne ocean sensors. IPY also saw the use of existing tools in new ways and in new places. These new tools allowed for a more comprehensive observation of the poles than ever before. The use of remotely controlled autonomous observing systems became increasingly common, while the cost and complexity of these systems often made multiagency and/or international cooperation necessary. This was never more apparent than with satellite systems. IPY cannot claim credit for the generation of any new satellite missions, but it did succeed in an unprecedented set of coordinated observation from spaceborne sensors operated by multiple national space agencies. Through the IPY's Space Task Group, this Polar Snapshot was so successful that the group has remained and continues to cooperate with national space agencies for observations intended to maintain an effective space-based monitoring of the polar regions to help overcome what is a decreasing observational capability as many satellite systems age and fail.

Observations are of little value if they are not available to researchers. However, the challenges to availability multiply as the data volumes increase and the needs of interdisciplinary research extend to data of unfamiliar form and content. A number of existing data centers in the United States stepped up to this challenge, making data management expertise available to IPY projects and following through with mechanisms to receive, organize, store, and make available metadata of all types that would assist researchers in locating data relevant to a wide range of scientific pursuits.

# 5

# Knowledge to Action

IPY 2007-2008 occurred during a period of change in the U.S. research enterprise, as evidenced by an increase in and funding for two-way connections between knowledge and action—knowledge informing action, and action influencing the pursuit of knowledge. Thus scientists have been integrated as advisors in policymaking processes, and policymakers, local agencies and communities, and other stakeholders have been included in the initial design of problem-oriented research.

One factor fostering connections between knowledge and action was the 2002 National Science Foundation (NSF) requirement that all proposals address the "broader impacts criterion" (Box 5.1).

In keeping with this new imperative, the IPY Vision Report (NRC, 2004) called for "improv[ing] predictions" and improving understanding of social processes, in particular those "that shape the resilience and sustainability of circumpolar human societies." In addition, the IPY Joint Committee (JC) and International Programme Office (IPO) required that all JC-endorsed international projects describe plans "for addressing the education, outreach, and communication issues outlined in the Framework document" (Rapley and Bell, 2004). As noted in that Framework document, "IPY 2007-2008 aims to inform both governmental and scientific decision-makers, including funding and resource managers, on the roles and importance of polar regions."

The inclusion of social and human sciences in the IPY program and its increased focus on polar residents, including indigenous peoples, was a first in the 125-year history of IPY/IGY and a critical factor in shaping the IPY agenda toward more "applied" (knowledge-to-action) outcomes. Another reason for the focus on polar communities was the extent of recent environmental change in the polar regions—they are experiencing climate forcing, climate effects, and climate change response more significantly than elsewhere. These changes, many of which exceed the range of historical measurements, have underscored the very present reality of climate change and driven home the need for adaptation planning and mitigation of environmental impacts.

In recognition of the rapid changes in the Arctic environment, residents, state and federal land managers, and industry representatives have called on the scientific community to help inform decisions about adapting to a rapidly changing environment.[1] Infrastructure planners in coastal communities, for example, need reliable projections of polar influences such as the impacts of glacier and ice sheet mass loss on sea level rise and of the warming Arctic on continental winter weather patterns.

The Arctic receives greater attention because of the significance of its changes for the people who live and work in the region, and the pace of change in most of Antarctica differs in important respects from that in the Arctic. But in both cases endeavors that entail multiple-decade planning must factor in the challenges of a changing baseline, and changes in both places can have important impacts on the entire globe.

---

[1] See, for example, *http://ine.uaf.edu/accap/research/cross_region_dialogue.htm*; ACCAP (2010); and Lovecraft and Eicken (2011).

**BOX 5.1**
**NSF Broader Impacts Criterion**

In every proposal seeking research funding from NSF, the principal investigator must spell out the research questions, intended methods, and proposed budget for the project. Starting in 2002, researchers were challenged to think more broadly about the societal impacts of their proposed activity, guided by the following questions:

- How well does the activity advance discovery and understanding while promoting teaching, training, and learning?
- How well does the proposed activity broaden the participation of underrepresented groups (e.g., based on gender, ethnicity, disability, geography)?
- To what extent will it enhance the infrastructure for research and education, such as facilities, instrumentation, networks, and partnerships?
- Will the results be disseminated broadly to enhance scientific and technological understanding?
- What may be the benefits of the proposed activity to society?

SOURCE: NSF, 2007.

Recent and ongoing studies of the stability of the West Antarctic ice sheet and grounding line (e.g., Jenkins et al., 2010; Velicogna, 2009)—otherwise highlighted in Chapter 3—show that large changes to global sea level rise are possible in response to global warming. Information about polar changes is thus relevant to decisions that affect the lives of millions of nonpolar residents.

IPY addressed the growing concerns about polar changes by organizing public and educational forums, and IPY outreach efforts led policymakers to turn to polar scientists to help inform decisions. In turn, through their experience preparing presentations for and answering questions from general audiences, many polar scientists learned about the concerns of the public, educators, and stakeholders, and in some cases adapted their research based on this new knowledge. The result was an increase during IPY in basic research that considered possible applications and stakeholder guidance (e.g., Stokes, 1997).

The committee notes that "action" was not defined as a major goal by the IPY Planning Group (2003-2004), the JC, or the National Research Council specifically. For example, the 2004 NRC Vision Report does not refer to the creation of adaptation plans or mitigation policy. The interest in connecting knowledge with action emerged as a logical extension of research projects or in response to information needs associated with expanding human activities, particularly in the Arctic.

The applications and observations described below are presented as examples and are not a comprehensive review of the portfolio of IPY knowledge-to-action activities.

## KNOWLEDGE-TO-ACTION EXAMPLES

IPY gave impetus for increased interaction and discussions between scientists and practitioners, including community planning groups, as the practitioners sought relevant, science-based information as a basis for their planning. Because of the larger environmental changes and greater populations in the North, most of the applications noted below are from the Arctic region.

### Predictions, Projections, Forecasts, and Scenarios

IPY had a strong focus on observations and modeling to improve predictive capability, in part due to the need to understand and project the forcing and implications of cryospheric changes. The foundation for this focus was laid years earlier under the scientific-community-inspired "Study of Environmental Arctic Change" (SEARCH).[2] SEARCH eventually became in interagency initiative with an international legacy through the International Study of Arctic Change (ISAC), which was established in 2003. The SEARCH tripartite charge "Observing Change, Understanding Change, and Responding to Change" is explicit about informing action.

The record sea ice minimum in 2007, the first year of IPY, stimulated concerted efforts to understand its cause, project plausible future trajectories, and consider systemwide implications. Cooperative oceanographic cruises and remote sensing imagery provided by many nations in concert with sophisticated modeling studies provided a comprehensive picture of its shrinking extent and thickness.

With the rate of change in the Arctic outpacing traditional modes of scientific communication, the

[2] *www.arcus.org/search/sciencecoordination/development.php.*

international sea ice research community has also made progress in exploring innovative approaches of synthesizing observations of the ice cover and model simulations to track and project the evolution of the ice cover on the seasonal scale. Advances were made in the area of seasonal predictions of sea ice conditions, particularly through the use of ensemble approaches with coupled ice-ocean models (e.g., Zhang et al., 2008). The recognized importance of improved seasonal predictions led to development of the Sea Ice Outlook[3] in 2008 by the SEARCH community. It has brought together a diverse international group of leaders in the field of sea ice modeling and forecasting to share and discuss yearly predictions of the summer sea ice minimum beginning in early summer each year. The Sea Ice Outlook is widely viewed as one of IPY's key legacies (Calder et al., 2011; Box 5.2).

Just as sea ice loss resulted in mobilizations of the sea ice science community, melting glaciers also mobilized the glacier science community to understand processes and make better projections of future sea level rise. One of the most significant outcomes of IPY ice sheet research is the multisensory documentation of a net loss of ice from both the Greenland ice sheet and Antarctica with a corresponding increase in sea level. Investigators have tracked the melt areas on the Greenland ice sheet through a distinct melt signature in the passive microwave satellite data (SMMR and SSM/I), showing a significant increase in the melt of the ice sheet over the last 29 years with 2007 having the highest melt extent on record. This increase in melt is important directly to shrinkage of the Greenland ice sheet and sea level rise but also contributes indirectly by providing more melt water to lubricate the interface between the ice and bedrock on which it rests, causing the ice to flow faster toward the sea. The increase in glacial melt has clear and very significant implications for a variety of social, economic, and ecological management considerations. Sea level rise projections for this century are now 0.62 to 1.8 m (NRC, 2010a). This is a significant increase over the IPCC (2007b) estimates of 0.18 to 0.59 m, which did not include dynamic aspect of glaciers. Extensive civil actions are required to prepare for this reshaping of the world's coastlines.

[3] *www.arcus.org/search/seaiceoutlook/index.php.*

IPY-related predictive modeling has and will continue to play a role in helping companies, individuals, and governments assess various risks associated with changing ice conditions, sea level rise, permafrost degradation, and other effects of polar warming. Such assessments help inform a wide variety of decisions involving siting and insurance of property and infrastructure, community planning and zoning, construction of ice roads, emergency preparedness and disaster response, and long-term planning for moving military, industrial, and public infrastructure (and in some cases whole villages) to higher ground. See Box 5.3. Overall, the ability of ecosystems and human communities to adapt to the rapid changes under way at both poles related to global warming depends in large measure on how "healthy" those human and natural communities are at the outset. In natural ecosystems, resilience and adaptive capacity are related to species and trophic diversity. For human communities, resilience appears strongly related to the strength of human social networks and institutions.

### Information for Subsistence Communities in the Arctic

In part, groundwork for two-way communication was laid through the process of preparing the Arctic Climate Impact Assessment (ACIA, 2005). Led by the United States, this report detailed the myriad and wide-ranging impacts of climate change and gave governments of many nations justification to initiate programs to detect and document a changing climate. The voices of polar residents recorded prior to and during IPY expressed the message of urgency and the call for action. Several IPY studies undertaken in human health, community vulnerability, food security, and local observations of change, were intrinsically aimed at practical applications to be shared with polar communities, local agencies, and grassroots organizations.

IPY promoted the practice of returning usable data to communities to encourage and solidify the involvement of local people in research, long-term environmental monitoring, and heritage preservation. Specific examples include:

- New data sets were created by indigenous IPY participants and managed by a special IPY project, ELOKA (Exchange of Local Knowledge in the Arctic);

## BOX 5.2
## The Sea Ice Outlook

The Sea Ice Outlook— begun during IPY—is an international effort to provide annual Arctic sea ice forecasts. This effort has continued beyond the period of IPY and has promoted advances in sea ice prediction, integration, and coordination of ground-based observations, as well as provided a more complete picture of the predictability of the Arctic ice cover on different scales. The outlook has been able to draw in different stakeholder communities, including Arctic residents, federal agencies, nongovernmental organizations, and industry. It has thus fostered a community of practice that can take the next steps in operationalizing different ice prediction approaches. Sea ice predictions are necessary to plan, prioritize, and map out activities and requisite infrastructure in newly opened waters. The United States and other governments rely on sea ice predictions (Figure) to evaluate future needs for services (e.g., port facilities, search and rescue capability, oil spill response facilities), regulation (e.g., discharge controls, requirements for ice-capable ships, limits on fishing), and information (e.g., new mapping and charting, communications capability, research priorities, etc.). According to a survey of Sea Ice Outlook users.[a] it provides a bigger picture for looking at the data, planning for shoreline changes, projecting impacts on marine mammals, and understanding where the uncertainties lie in the forecasts.

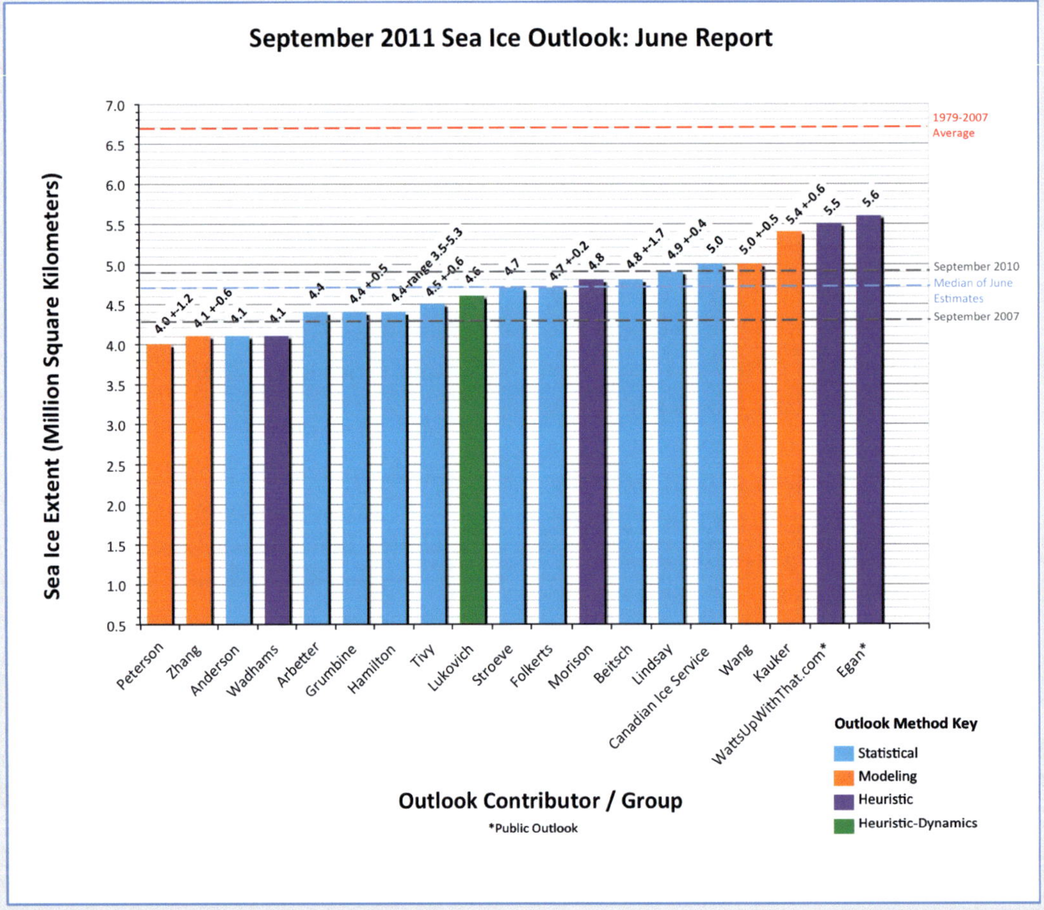


Predictions of the annual minimum arctic sea ice areal extent, compiled from numerous sources, are provided monthly (June through October) as a forum to facilitate communication among sea ice researchers and provide information to stakeholders, including Arctic residents, federal agencies, nongovernmental organizations, and industry. SOURCE: Study of Environmental Arctic Change/Arctic Research Consortium of the United States.

[a] *www.arcus.org/search/seaiceoutlook/survey.*

**BOX 5.3**
**Climate Change Adaptation**

Climate change is already having measureable effects on the polar regions. Whether beneficial or detrimental, those effects will require ecosystems and human communities to adapt to the changes. Global climate models are advancing our ability to understand what changes are coming and beginning to help inform decision about how to adapt to potential impacts. Important examples are the models included in the Intergovernmental Panel on Climate Change (IPCC) assessment process. IPY came at an important time when the IPCC models were being developed in preparation for the Fifth Assessment Report (AR5). IPY produced a wealth of information that will be featured prominently in several chapters of AR5 (in preparation), in particular in a chapter on "Polar Regions" in the volume on *Impacts, Adaptation, and Vulnerability* (Working Group II). Other modeling work has shown the influence of climate forcings at various latitudes, emphasizing that forcings at the midlatitudes in the Northern Hemisphere strongly influence Arctic temperatures (Shindell et al., 2010).

- A new Web-based monitoring and data-sharing network in the Bering Strait region, Sea Ice for Walrus Outlook SIWO[4] (Eicken et al., 2011), was developed by ice scientists in partnership with the Eskimo Walrus Commission and several local village monitors; SIWO uses high-resolution satellite images, analysis of weather and ice patterns, and observations from local scientists and indigenous experts to provide forecasts for the spring ice breakup and the walrus migration in the northern Bering Sea region in a format that is helpful to local users, as well as regional 10-day weather forecasts;
- A long-term study of ice trails built by indigenous whalers across spring shore-fast ice off Barrow and other Alaskan communities shared digital maps of trails with the community (SIZONeT project); and
- Various efforts to share computer and Web-based maps and satellite imagery of subsistence sea ice use, hunters' and herders' traveling, local impact of oil and gas development, marine mammal distribution, and other information were part of IPY.

Risk assessments done during IPY have also played an important role in predicting coastal erosion, loss of permafrost, and other changes faced by subsistence communities around the Arctic. For example, a valuable product of the IPY period was an international assessment of Arctic coastal erosion (IASC, 2011). This collaborative research advanced the understanding of the processes responsible for the recent increase in erosion rates throughout the circumpolar Arctic, but more importantly, it provided scientific insight needed to evaluate areas of stability and areas of vulnerability. Throughout the Arctic, many small villages and towns are located adjacent to the coast or on rivers. With accelerated permafrost thawing, loss of sea ice armor, and increase in summer storms, many coastal communities now face imminent threat of erosion and possible destruction. In Alaska, the villages of Shishmaref, Kivalina, and Newtok have already begun relocation plans. Since 2003, federal, state, and village officials have identified 31 villages that face imminent threats from flooding and erosion. The U.S. Army Corps of Engineers has identified over 160 additional rural communities threatened by erosion (GAO, 2009).

The Arctic Human Health Initiative (AHHI) was the U.S.-led IPY coordinating project introduced via the Arctic Council with the overall goal to increase awareness and visibility of human health concerns of Arctic peoples, foster human health research, and promote health strategies to improve health and well-being of Arctic residents. It was a broad circumpolar effort with multinational participation that included almost 30 individual projects in several thematic fields: health network expansion, infectious disease research, environmental health, and behavioral and mental health (Parkinson, 2010). Among many U.S. contributions, the Center for Alaska Native Health Research (CANHR) at the University of Alaska, Fairbanks used the IPY momentum to build a collaborative research presence in Alaska Native communities, focusing on prevention and reduction of health disparities—particularly in the areas of behavioral, dietary, and genetic risks—and protective factors related to obesity, diabetes, and cardiovascular disease risk in Alaska Natives. All CANHR studies, particularly those related to substance abuse and suicide prevention, the development of novel dietary biomarkers, contaminants, and the safety of subsistence foods, employed community-based participatory research approaches. Also, during IPY, opportunities were created for cross-border

[4] *http://www.arcus.org/search/siwo.*

partnerships to explore needs related to service delivery. Together, the NSF and Alaska Federal Health Care Access Network (AFHCAN) facilitated cooperation in telemedicine technology expertise between Alaska and the Sakha Republic and the Khanty-Mansiysk region in Russia. The goal of this partnership was to promote a mutually beneficial collaboration in telemedicine, telehealth, mobile medicine, and distance learning in remote areas of Alaska and the Russian north.

The IPY project entitled Arctic Change: An Interdisciplinary Dialog Between the Academy, Northern Peoples and Policy Makers led to several diverse workshops to capture talking points important for informing policymakers on issues of Arctic security and climate change, Arctic health, Arctic Ocean shipping, and human security in a changing Arctic. The Arctic Institute for Applied Circumpolar Policy (IACP[5]) was founded as an outgrowth of this framework. IACP is a joint effort of Dartmouth College, the University of Alaska, and the University of the Arctic and brings together representatives of governments, the academy, nongovernmental groups, and indigenous peoples to discuss Arctic and polar issues, identify and prioritize the policy-related research requirements, and help develop the agendas for governments to address pressing policy issues facing the northern and polar regions.

The inclusion of social and human-focused research in the IPY program is broadly viewed as one of its key features as well as major achievements. It transformed into a massive flow of new knowledge that produced tangible benefits to many stakeholders beyond participating scientists. It included research efforts supported by a broad spectrum of governmental agencies, such as NSF, the National Park Service (NPS), National Oceanic and Atmospheric Administration (NOAA), U.S. Department of Health and Human Services (HHS), and the Smithsonian Institution, as well as the University of Alaska, North Pacific Research Board, and other public and private players. The overall U.S. IPY input in social and human fields was unprecedented in its scope and funding (more than $20M in the United States alone over 3 years) and resulted in the largest and the most diverse effort of its kind.

[5] *http://iacp.dartmouth.edu/.*

## Information for Shipping

Decline of sea ice during the summer months will create increased access to large and small ships from subarctic as well as Arctic nations; for example, the Arctic Ocean will become the shortest shipping route between Hong Kong and New York. The Arctic marine and terrestrial environments are fragile, and oil spills or other environmental disruptions will be more difficult and take longer to remediate than those in the lower latitudes. Sea ice loss projections related to IPY inform many different decisions regarding shipping routes, port siting, emergency response (see below), ship construction, and others.

The Arctic Marine Shipping Assessment (Arctic Council, 2009), prepared during IPY, concluded that the most significant threat from ships to the Arctic marine environment is the release of oil through accidental or illegal discharge. That potential problem is compounded by the lack in the Arctic of emergency response capability for saving lives and mitigating pollution. This assessment spurred an agreement to negotiate a new mandatory Polar Shipping Code.

In the Antarctic, growing awareness of polar ecosystems and biodiversity, along with burgeoning commercial activities in the region, sparked new interest in tighter controls on shipping and fishing vessel operations and upgrading emergency response capability in the region. Tourist and fishing vessels gravitate toward areas of high diversity or productivity, and their activities can be inconsistent with maintaining both. This was highlighted when the MV *Explorer* was sunk by ice in November 2007 (Figure 5.1). Since that time, the International Maritime Organization has extended the Polar Shipping Code to waters at both poles.

## Information for Emergency Preparedness

As shipping, energy development, tourism, and other human activities expand throughout the Arctic, awareness of the need for effective emergency planning has grown. For example, increasing interest in the polar regions as tourist destinations reveals the risks inherent in traveling in remote and sometimes dangerous locations. For example, when the tourist ship *Explorer* capsized near the Antarctic Peninsula in 2007 (Figure 5.1, top), passengers had to rely on other nearby

FIGURE 5.1 Top: The tourist ship *Explorer* capsized near the Antarctic Peninsula in November 2007. SOURCE: Fuerza Aerea de Chile via European Pressphoto Agency; Bottom: The Antarctic tourist ship *Clelia II* without power and making slow headway north of the South Shetland Islands in rough seas during December 2010. SOURCE: Copyright Stewart/McIntosh

cruise ships. More recently, the tourist ship Clelia II lost power and communications during particularly rough seas in the Drake Passage (Figure 5.1, bottom). Projections of sea ice changes during IPY have propelled and informed emergency preparedness and response mechanismsnd will continue to do so. Effective emergency preparedness and response will have to draw on data from both research and operational observing systems developed during IPY, with high demands placed on data availability and spatiotemporal resolution during an emergency. Hence, such information needs may turn into a powerful driver outside of the research community towards collaborative, internationally coordinated activities governed by open data/access practices as promoted during IPY.

As part of the Arctic Marine Shipping Assessment, the Coast Guard, the U.S. Arctic Research Commission, and the Coastal Response Research Center (a partnership between the University of New Hampshire and NOAA) organized a workshop of international experts to anticipate responses to environmental and safety incidents in the Arctic (Coastal Response Research Center, 2009). The workshop identified gaps in current response capabilities, assessed future response needs, and recommended improvements in the ability of Arctic nations and indigenous communities to prepare for and respond to marine incidents.

In response to the growing recognition of the need for more effective disaster planning, discussions were initiated during IPY that led to an agreement on search and rescue coordination that was signed in May 2011 by the foreign ministers of the eight Arctic states that constitute the Arctic Council.

## Information for Ecosystem Management

IPY identified new marine and terrestrial species, habitats, and ranges, which greatly expanded understanding and awareness of polar biodiversity. Better understanding of polar ecosystem dynamics has in turn spurred a number of new initiatives aimed at managing human activities in the oceans, with an eye toward protecting biodiversity and maintaining ecosystem functions as these ecosystems undergo profound change due to global warming, including ocean acidification (Box 5.4). For example, the Arctic and ecosystem-based management are among nine strategic priorities identified under the United States' newly adopted U.S. National Oceans Policy. Since the July 2010 Executive Order establishing the Policy, the Administration has moved forward with a number of initiatives to advance ecosystem-based management in the Arctic Ocean. In May 2010 the foreign ministers of the eight Arctic Council states agreed to establish an expert working group on ecosystem-based management. Ecosystem and species mapping and predictive modeling is helping to identify ecologically important and vulnerable areas, and will help inform new processes initiated within the United States and at the Arctic Council to promote ecosystem-based marine resource management.

> **BOX 5.4**
> **Ocean Acidification**
>
> The increase in seawater acidity (decrease in pH) due to the uptake of anthropogenic carbon dioxide has been termed "ocean acidification." Given the scenarios for pH changes in the Arctic Ocean and adjacent Arctic shelves seas, there will likely be an increasing impact by ocean acidification, with potentially negative implications for shelled benthic organisms as well as those animals that rely on the shelf seafloor ecosystem with consequent impacts to the fishing industry such as crab fishing. (Also see section in Chapter 2 on "Marine Carbon Cycling and Ocean Acidification".)

After extensive debate and analysis of the consequences of the loss of sea ice on polar bear habitat, the U.S. Fish and Wildlife Service determined that a viable threat exists and will continue to threaten the polar bear species. A ruling was published in the *Federal Register* (U.S. Department of the Interior, 2008) on May 15, 2008, listing the polar bear as a threatened species under the Endangered Species Act (ESA). The ruling found that changes in the abundance, distribution, or existence of sea ice will have effects on the number and behavior of these animals and their prey.

## Information for Fisheries

Documentation of the northward movement of commercial fish populations that occurred during IPY has opened the possibility of new commercial fisheries that will need management (at present, there is no international management mechanism in place to manage fisheries in most of the Arctic Ocean). This possibility led the United States to proactively prohibit most commercial fishing in its waters north of the Bering Strait until better scientific information is available on the ecology of the region and the impacts of new commercial fishing on both subsistence users and the marine environment. As one online questionnaire respondent stated as part of the input to this report, "gathered biological data are providing the first pan-Arctic baseline to assess changes in Arctic marine biodiversity." Informal international discussions are also under way regarding the establishment of some mechanism to manage commercial fisheries in the area of the Arctic Ocean where such mechanisms are lacking.

In the Antarctic, IPY-related scientific data are being generated for use in policy-relevant conservation and management efforts related to fisheries and tourism. The Conservation of Antarctic Marine Living Resources (CCAMLR) is an organization that was established in 1982 as part of the Antarctic Treaty system. CCAMLR manages and sets fishing limits in the Southern Ocean, identifies needed research, and is involved in monitoring environmental impact. The Southern Ocean Global Ocean Ecosystems Dynamics Program (SO GLOBEC; described in Chapter 4), was an international multidisciplinary effort during IPY, designed to examine the growth, reproduction, recruitment, and overwintering survival of Antarctic krill (*Euphausia superba*). The rising recognition of krill as a key element of Antarctic ecosystem function during IPY led CCAMLR to spatially allocate the fishery to prevent the catch from being concentrated in a small area, and to mandate scientific observers on at least half the ships harvesting krill.

## Information for Offshore Oil and Gas Development

In 2008, the U.S. Geological Survey (USGS) completed the Circum-Arctic Resource Appraisal, an assessment of undiscovered conventional oil and gas resources in all areas north of the Arctic Circle (Figure 5.2). The USGS estimated that the Arctic accounts for about 13 percent of the undiscovered oil, 30 percent of the undiscovered natural gas, and 20 percent of the undiscovered natural gas liquids in the world, about 84 percent of which are expected to occur offshore. It is estimated that these resources may account for about 22 percent of the undiscovered, technically recoverable resources in the world.

Planning for Arctic offshore oil and gas development requires projections of sea ice as well as other marine ecosystem information, from subsea permafrost status and trends to the projected distribution of marine wildlife in and around drilling sites. Information generated during and after IPY will play an essential role in permitting and other decisions (National Commission on the BP Deepwater Horizon Oil Spill and Offshore Drilling, 2011).

## Information for Onshore Development

Arctic and subarctic wildfire frequency and severity have increased markedly in the past decade (Kasischke et al., 2011) with important impacts to natural ecosystems and the social network dependent upon them (Chapin and Lovecraft, 2011). Insurance companies are revisiting their procedures for coverage of structures in fire-prone areas, including prescribing techniques of fire protection to home owners. State, federal, and local land managers need to strategically position equipment and manpower and to consider long-term effects of fire and the changing trajectories of ecosystem recovery (Payne, 2010).

Construction and new development in a period of warming present unprecedented challenges for design engineers. In the past, design protocols were based upon compilations of measurements of actual field conditions (for example, Hartman and Johnson, 1978). However, documentation of circumpolar warming has forced engineers to realize that future environmental conditions will fall outside the domain of historical observations; therefore it is necessary to include projections of warming and potential thawing in design of roads, buildings, and other infrastructure (McGregor, 2010).

Projects such as the IPY Thermal State of Permafrost (Romanovsky et al., 2008) demonstrated to the engineering community that warming is nearly ubiquitous throughout the high northern latitudes and that new approaches are required for construction in ice-rich environments. "Warming rates are much smaller for permafrost already at temperatures close to 0°C compared with colder permafrost, especially for

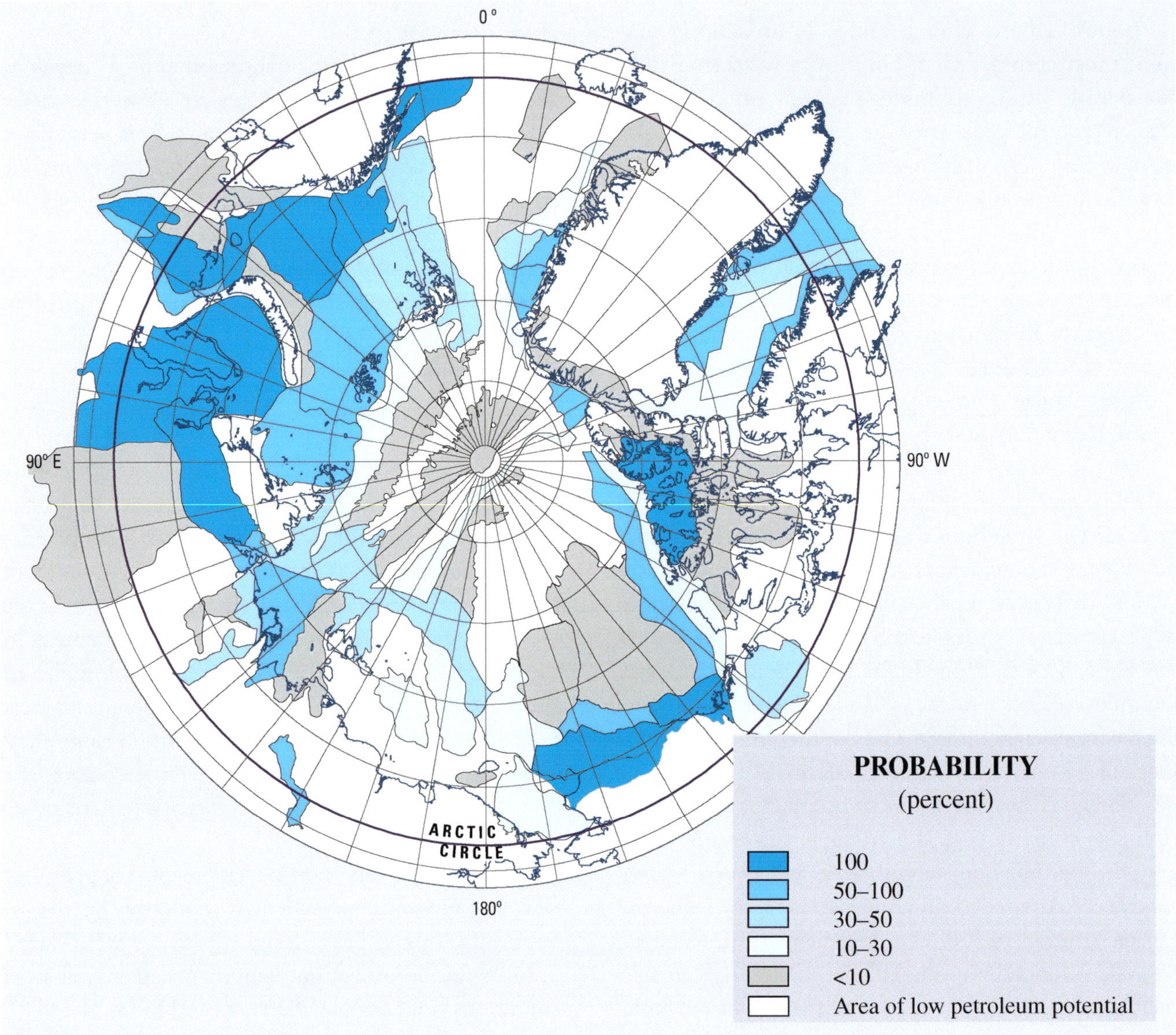


FIGURE 5.2 Potential oil and natural gas reservoirs in the Arctic account for a significant percentage of the world's energy resources. Map shows assessment units (AUs)—mappable volumes of rock with common geologic traits—in the Circum-Arctic Resource Appraisal (CARA) color-coded by assessed probability of the presence of at least one undiscovered oil and/or gas field with recoverable resources greater than 50 million barrels of oil equivalent (MMBOE). Probabilities for AUs are based on the entire area of the AU, including any parts south of the Arctic Circle. SOURCE: Bird et al., 2008.

ice-rich permafrost where latent heat effects dominate the ground thermal regime. Colder permafrost sites are warming more rapidly."[6] New construction techniques are being developed to passively cool roadbeds in ice-rich permafrost terrain to maintain thermal stability and structural integrity (Xu and Goering, 2008).

Continued warming will limit use of ice roads, which are now commonly used in Northern Canada, Alaska, and Siberia. By 2050, inland regions of the Arctic, now accessible through seasonally constructed ice roads, may become inaccessible (Stephenson et al., 2011). Presently, seasonal transportation and construction in winter allows access to vast roadless regions, minimizing environmental impacts and construction costs. Most of the communities connected by the roads

[6] Vladimir Romanovsky, University of Alaska, Fairbanks, personal communication, 2011.

are small, remote villages, so it will not be economical to replace the ice roads with all-weather roads.

### Issues of Sovereignty and Security

Rapid environmental changes establish conditions that create more favorable local climates in some regions of the Earth but that in other regions may threaten societies and the environment. Even in the midlatitudes, climate change will exacerbate drought in some regions and flooding in others, enough to make human habitation difficult in some areas where cities currently flourish (IPCC, 2011). The U.S. Army has realized the potential of this to cause international conflict, and this has been an increasing topic of concern and activity in U.S. Army future scenario planning.[7]

Seabed mapping and sampling associated with IPY has been important to inform the development of the U.S. submission to the Commission on the Limits of the Continental Shelf regarding territorial claims on the outer continental shelf. Understanding of the continental shelf and its relationship to the surrounding seas is required under the Law of the Sea Treaty, which allows the Arctic rim countries, including the United States, Canada, Denmark, Norway, and Russia to make claims for undersea resources on, above, and underneath the seabed up to 200 miles from their natural coasts.[8]

In the summer of 2007, just as IPY was getting under way, a Russian expedition sent a submarine to the seabed on the Lomonosov Ridge to plant the Russian flag, claiming that it was an extension of their continental shelf. This action did not trigger any response or reaction from the U.S. military. As reported by a think tank of international representatives from government, military, and economic sectors, there is no perceived military threat in the Arctic. Rather, the greatest security threat in the Arctic arises from environmental or natural disasters, and an urgent need remains to establish regional and international coordination and cooperation in preventing and mitigating such events (Carnegie Foundation, 2008).

The U.S. Navy Postgraduate School in Monterey, California, was active in sea ice modeling and monitoring, though in contrast to the IGY, the operational Navy was not an active participant in IPY 2007-2008.[9] However, as the importance and urgency of the Arctic ice retreat became evident during the IPY years, the Navy, U.S. Coast Guard, NOAA, ONR, National Ice Center, and NSF cosponsored three symposia on the "Impact of the Ice-Diminishing Arctic on the Naval and Maritime Operations."[10] Also, in 2011, ONR reestablished targeted research efforts in polar regions through the Arctic and Global Prediction Program. ONR will expand upon the extensive understanding of sea ice dynamics that was gained during IPY.

### Polar Policy

In response to the information generated in part during IPY, governments have undertaken significant revisions of their policies. While the United Nations Law of the Sea still remains unsigned by the U.S. government, there were assertive actions by U.S. governmental agencies during IPY. During IPY years, the projections for ice-free Arctic summers, rising sea level, and increased need for disaster response in the Arctic led to U.S. Navy and Coast Guard planning for Arctic conditions that are unlike those of past decades (Arctic Council, 2009). In January 2009, President George W. Bush signed the National Security Presidential Directive 66 and Homeland Security Presidential Directive 25 on Arctic Regional Policy. These directives establish policies aimed at meeting national and homeland security needs in the Arctic, protect the Arctic environment and conserve its biological resources, strengthen institutions for international cooperation, involve the Arctic's indigenous communities in decisions that affect them, and enhance scientific monitoring and research in local, regional, and global environmental issues.

IPY also coincided with a general recognition that Arctic geopolitics have entered into a new era of strengthened indigenous rights, increased attention

[7] *www.armyscienceconference.com/.*

[8] *http://oceanservice.noaa.gov/facts/eez.html.*

[9] In 1948, the US Navy, Office of Naval Research (ONR) established the Naval Arctic Research Laboratory (NARL) in Barrow, Alaska. With the end of the Cold War and no broad acceptance that climate change was a serious problem, this facility was closed 1980 and ORN focused on research in more temperate regions. In 2000, the US Navy began to consider the actions needed to prepare for Naval Operations in an Ice-free Arctic.

[10] In 2007, *www.star.nesdis.noaa.gov/star/IceSymposium.php*; 2009, *www.star.nesdis.noaa.gov/star/IceSymposium2009.php*; 2011, *www.star.nesdis.noaa.gov/star/Ice2011.php*.

to Arctic matters by non-Arctic Asian and European nations, and an emerging role of the Arctic Council and other international frameworks. These trends were manifested through increasingly active role of the Arctic Council and its six Permanent Participants representing polar indigenous peoples in promoting new science initiatives during the IPY years and in promoting IPY itself. Several IPY projects and resulting publications addressed the sociopolitical and policy aspects of the changing status of the polar regions (i.e., Berkman et al., 2011; Launius et al., 2010; Shadian and Tennberg, 2009).

The Inuit Circumpolar Council (ICC) is an international indigenous peoples' organization representing approximately 160,000 Inuit living in the Arctic regions of Alaska, Canada, Greenland, and Chukotka (Russia). The ICC had U.S. partners at universities and agencies within the United States during IPY, both for actions on joint IPY projects and also for discussions on international issues. In 2007, the ICC was successful in moving forward a new United Nations declaration on the Rights of Indigenous Peoples Act.[11] UN Resolution 61/295 adopted in September 2007 identifies the declaration as an international standard of achievement to be pursued in a spirit of partnership and mutual respect, with 46 articles spanning a large range of rights, including self-determination, rights not to be subjected to forced assimilation or destruction of culture, rights to participation in decision-making matters that would affect their rights, and rights to redress for lands and traditional resources that have been used or damaged without their consent. As perhaps the first people to be severely affected by a climate change caused primarily by industrialized nations, the Inuit moved to action during the IPY years.

In 2008, ICC convened an IPY climate change policy workshop aboard the vessel CCGS *Amundsen*, which brought together climate change scientists and Inuit leaders to address the effects of climate change in the Arctic region. Based on insights from these leaders, the ICC released the "Amundsen Statement: 2012 Climate Change Roadmap,"[12] which highlighted their strategy for addressing the potential impacts of global climate change. Building upon the Amundsen statement and in response to rising greenhouse gas emissions and the devastating effects of warming in the Arctic, the ICC issued a Call to Action during the COP15 meetings that addressed many issues, including calling on global leaders at COP15 to help sustain Inuit lands and territories by ratifying a post-2012 agreement to help stabilize greenhouse gas concentrations at 350 ppm in order to maintain long-term global temperature increases well below 2°C.

While the Arctic had the Arctic Climate Impact Assessment (ACIA, 2005) as a foundation coming into IPY, an analogous report was developed and released for the Antarctic during IPY. Called the Antarctic Climate Change and the Environment (ACCE; Turner et al., 2009), the report was sponsored by the Scientific Committee on Antarctic Research (SCAR). Like ACIA, it is a seminal synthesis of scientific findings that provides a comprehensive and authoritative analysis useful for informing policy decisions.

The Antarctic Treaty (Secretariat of the Antarctic Treaty, 2009), a major international document that followed on the heels of the IGY years, has been periodically updated, and a variety of new initiatives that link policy with the environment were advanced at several Antarctic Treaty Consultative Meetings during the IPY time frame, specifically the ATCM XXXII in Baltimore, Maryland (April 2009), which was the first-ever joint meeting of the Arctic Council with the Antarctic Treaty Consultative Meeting, and the Antarctic Treaty "Summit" dedicated to the 50th anniversary of the Antarctic Treaty in December 2009 (Berkman, 1960). These initiatives were informed by the ACCE report and other findings and included a number of science-based conservation and protection initiatives put forth by the Committee on Environmental Protection on a broad spectrum of topics, including climate impacts on the environment, biological indicators of human impact, disturbance on wildlife, and introduction of nonnative species. The ACCE had multiple U.S. coauthors, and the report has had wide visibility in Antarctic Treaty meetings, in IPCC discussions, and the United Nations Framework Conventions on Climate Change (UNFCCC) meetings.

[11] *www.inuit.org/index.php?id=267.*

[12] *www.inuitcircumpolar.com.*

## Decision Making Beyond the Poles

### *Global Sea Level Rise*

The current rapid climate change that is most evident in the polar regions will affect all of humanity, either directly or indirectly. The ongoing demise of glaciers and ice sheets is contributing to global sea level rise that affects coastal communities and cities worldwide, but due to gravitational effects, it will cause the most marked rise in North America and the Arctic (Raymo et al., 2011). Research conducted during IPY helped quantify how the ice sheets are changing, advanced our understanding of the driving mechanisms, and furthered our knowledge of the rheology of the ice sheets. Flooding of cities and increased coastal storm damage is projected to cause billions of dollars of damage within the timespan of a human lifetime (IPCC, 2007a; Nicholls et al., 2007).

Public awareness of the role of ice sheets and glaciers in the climate system and their direct effect on sea level grew during IPY. For example, the California Coastal Commission met with scientists in 2009 with the aim of using scientific results on sea level rise to inform their decisions on infrastructure planning along the California coast.[13] The state of Delaware has developed a sea level rise action plan that rests upon estimates of future sea level rise from scientific studies (Valencik, 2010). New York City's "Responding to Climate Change in New York State" assessment running from 2008-2010 (Rosenzweig et al., 2011) used "rapid ice melt scenario based on accelerated melting of the Greenland and West Antarctic Ice Sheets" in its projection of potential sea level rise.

### *Sub-Arctic Weather*

The "Warm Arctic-Cold Continent" weather pattern can influence sub-Arctic weather and is thus important for midlatitude forecasts.[14] Characterizing and quantifying teleconnections among polar processes and subpolar or temperate region responses is difficult. Record warm weather in the Arctic over the past 5 years may have played some role in affecting the weather in lower latitudes, including colder winter temperatures. The character of extreme winter events is influenced by many factors, including both climate oscillations such as El Niño, as well as longer-term trends such as changes in the Arctic stratosphere and snow cover. Changing weather patterns in the midlatitudes, in some cases precipitated by changing conditions in the polar regions, will affect agriculture, forestry, and lifestyles in many places on Earth. These teleconnections of the Warm Arctic-Cold Continent require further research through observational and modeling studies (this is also described in Chapter 2 in the section on "Sea Ice Vulnerability and Teleconnections to Society").

## Levers and Hurdles

In making connections between knowledge and action, there are many hurdles that need to be overcome, with some interesting levers (opportunities to encourage action) identified through IPY activities.

The IPY framework served several important functions in connecting knowledge with action. IPY was seen as a neutral space where the goal was the pursuit and dissemination of knowledge. Academics and researchers were largely viewed as honest brokers who would represent their findings without bias. For example, the *Extreme Ice Survey*[15] documentary on the National Geographic Channel during IPY generated great interest and public engagement. This translation of climate information into public action is new and growing, and it was greatly facilitated by the publicity generated by IPY. Thus IPY contributed to a knowledge base that could then be used by others to suit their needs.

A new aspect of the "knowledge to action" function of science research during this IPY is that it helped engage local stakeholders and was instrumental in generating the capacity-building momentum in the U.S. polar regions. In earlier IPYs/IGY, there was little if any local research infrastructure in Alaska besides the fledging university campus in Fairbanks (then called "College"). During the IPY 2007-2008 era, numerous local players in the State of Alaska were among the key beneficiaries of the U.S. engagement in IPY. The University of Alaska system of three urban and several rural campuses ran its own IPY program that was one of the

---

13 *http://institute.lanl.gov/igpp/_docs/Final_SLR_workshop5-11.pdf.*

14 *www.arctic.noaa.gov/future/warm_arctic_cold_continent.html.*

15 *www.extremeicesurvey.org/.*

largest in the nation; this included active IPY research programs in a wide variety of scientific disciplines and the initiation of 11 postdoctoral fellowships for research that embraced the IPY philosophy and criteria.

The newly expanded Barrow Arctic Science Consortium facility in Barrow acted as a major hub for several IPY projects, ensuring the flow of resources, knowledge, and practices to the Barrow community and local institutions. State offices of many federal agencies, including NOAA, USGS, NPS, HHS, Fish and Wildlife Service, Environmental Protection Agency, and U.S. Coast Guard were actively engaged in IPY research. IPY produced tangible practical outcomes to local stakeholders, such as improved services, flow of data, improved data management, and monitoring capacities, as well as active outreach programs.

A special "knowledge to action" impact of IPY was the engagement of northern residents and indigenous organizations.

> The inclusion of "human dimensions" in IPY 2007-2008 program took it to the next level, but the vision of the IPY organizers eventually expanded the notion of inclusiveness to the range never experienced in the previous "polar years." Arctic residents, especially indigenous peoples, were recognized as important stakeholders, collaborators and drivers of new research, and, for the first time, were explicitly called upon to participate in IPY science. (IPCC, 2007a)

Other indigenous organizations in the State of Alaska (Eskimo Walrus Commission, Nanuq/Polar Bear Commission) as well as dozens of local communities, from Barrow to tiny Shaktoolik (population 160) took part in the impressive spectrum of IPY research, from sea ice and weather observations to language documentation, human health, and community heritage programs. New technologies, improved data management, and knowledge sharing (Gearheard et al., 2011), better forecasting and health services, trained local personnel, and science-inspired indigenous youth were the obvious benefits of the unprecedented local engagement in IPY research. Nothing of this kind had been achieved in any previous national polar program, including IGY 1957-1958 or the earlier IPY-1 and IPY-2.

As polar scientists were drawn during IPY into working with stakeholders, scientists faced several challenges. First, most polar scientists are not trained to provide actionable advice; they are trained to conduct and interpret scientific analyses. Second, through their experience with IPY, many participants became increasingly aware of the value of communication through education and outreach, but they also came to recognize that different audiences require different communication approaches. This has challenged the polar community to learn how to express themselves in ways that are scientifically accurate but also meaningful to a variety of audiences. Several IPY-supported activities invested in training scientists to be better communicators.

Third, anthropogenic contributions to the warming and acidification trends observed in polar regions have led many polar scientists to see the need to reduce future impacts through decreasing the emissions of greenhouse gases. As researchers seek, or are called on, to integrate their research with decision making, they find that they encounter a complex set of issues in connecting knowledge with action. Who makes decisions? What information do they need for their decision making? Who implements changes? What structures incentivize change? Interactions have become increasingly difficult with the polarization of the political and cultural landscape (Overland and Wang, 2009). This has further complicated the task many scientists faced in navigating the line between conducting polar research to *inform* action, and conducting research to *promote* action. While the community survey run by this committee revealed concerns when there is a "mixing of advocacy with science" (Survey Response #157392819), there is a growing sense in the polar research and education community that "IPY has made 'knowledge to action' a proper domain for scientists—in the past it was often disregarded as 'activism'" (Survey Response #158556187).

Lastly, with respect to scientific data, decision makers, educators, and the general public are increasingly accessing data directly in order to tailor their own analyses and interpretations. The fundamental concept of the Arctic Observing Network (AON) was rapid access to all data. Specifically, NSF, as a core supporter of AON, reflected decision maker and scientific community interests in stipulating that data from AON cannot be embargoed and needs to be made available immediately after collection. With more than 50 active AON projects, this open-access policy ensures that long-term Arctic observations collected through a research network can also help serve increasingly important operational needs.

**BOX 5.5**
**Perspectives on "Knowledge to Action" During IPY**

In the course of conducting this study, the Committee gathered perspectives on the importance of converting knowledge to action during IPY from those inside and outside of the polar research community. Some examples include:

"The U.S. has huge geopolitical interests in the Arctic region, and we need to understand the changes that are taking place there. Many other countries have direct economic interests in the Arctic, and all are served by joining forces in IPY research. Additionally the rapidly diminishing ice in the Arctic is creating new opportunities for transport and marine resource development."

— John H. Marburger III, U.S. Science Adviser under President George W. Bush (Revkin, 2007)

"I would think a strong take-away from this IPY would be a decision to include effective communications plans, and resources for them, in every aspect of the effort."

—Roger Launius, Smithsonian Institution

"Significantly, IPY has made impressive progress with the last goal—educating the public and decision-makers. As a variety of high profile events and publications shared some of the program's early scientific results, it became increasingly obvious that national and international policy-makers and the public are beginning to recognize the Arctic's scientific and strategic importance. From international diplomatic events to presidential policy changes and increased science budgets, the events of the past few months show that arctic science no longer operates in obscurity. In this new era of arctic awareness, it is incumbent on the members of the research community to be prepared—both to maximize the many opportunities the new era brings and to think through the policy implications of their work."

—Mead Treadwell, Lt. Governor of Alaska (Treadwell, 2009)

"Key challenges in the post-IPY phase are how to sustain this engagement, which will require an investment of some sort to help institutionalize the hubs that allow much of these activities to occur. Here, a concerted effort across different federal, state and local agencies is needed."

— Hajo Eicken, University of Alaska, Fairbanks

"Many elements of the IPY networks and initiatives can provide the seeds for further development of a comprehensive polar observing system. At present, these initiatives are acting separately, and their integration, ensuring data delivery and optimization should turn them into the first functioning polar observing system, which is able to provide data for scientific research and practical applications. Cooperation of the IPY-born observing systems mostly driven so far by scientific and academic institutions with agencies having operational responsibilities should be encouraged to sustain the achieved and required IPY legacy in terms of polar observations."

— Vladimir Romanovsky, University of Alaska, Fairbanks

"Over the past decade, including the IPY, the awareness of the scientific community has been heightened with respect to stakeholder needs. We are now paying much more attention to stakeholder needs and we are integrating these needs into our research planning."

— Peter Schlosser, Columbia University

"We are encouraged by discoveries made during the International Polar Year. Look at what's been accomplished: scientists produced detailed maps of the last unexplored mountain range on Earth, sent robot submarines under the Antarctic Ice Shelf to map the sea beds, drilled deep beneath the sea floor to learn more about the effects of carbon dioxide on the West Antarctic Ice Sheet, and shed light on how climate change affects the microscopic life at the base of our ecosystem. Together, these discoveries will advance our understanding and hopefully inspire us to work more closely together to limit the impacts on our lives."

— Hillary Clinton, U.S. Secretary of State (Clinton, 2009)

For example, in development of their "Arctic Roadmap," the U.S. Navy includes the Arctic Observing Network among the important assets and describes the need to increase operations of unmanned systems for Arctic data collection, monitoring, and research (U.S. Navy, 2009). Compared to the data sharing that occurs within academic research groups, there is comparatively modest engagement by the private sector. "In particular the resource industries are making major investments in the Arctic, yet, there is comparatively little coordination and data exchange between industry and academia."[16] There may be a role for regulatory agencies to promote free data access along with leasing requirements.

## CONCLUSIONS

The polar science community gained a wealth of experience during IPY in learning to understand and

[16] Hajo Eicken, University of Alaska Fairbanks, personal communication, 2011.

manage change in a time of change (Box 5.5). IPY showed that training and continued exposure of scientists to addressing real and present issues increases their effectiveness in connecting knowledge with action, by improving communication skills, influencing research agendas, and by direct experience of navigating the line between communication and advocacy. This sets a stage of readiness within the polar community that is important for tackling future research endeavors. Looking ahead, further development of two-way communication, observing systems, and predictive capability is needed to maintain and extend connecting knowledge with action. At the time of this report-writing, the concluding IPY Conference is scheduled to take place in April 2012, and will highlight this theme of "knowledge to action."

# 6

# Reflections

With a scientific focus on advancing understanding of the polar regions in a time of rapid planetary change, IPY 2007-2008 was the right initiative at the right time. Dramatic environmental changes were occurring in polar regions, and the research community was ready with the tools and the expertise to investigate these changes in sophisticated new ways.

The success of IPY can be attributed to its timeliness in addressing the key roles of the polar regions in the Earth system; its international reach, making it possible to exploit initiative and capabilities worldwide; its unprecedented breadth of interdisciplinary involvement, from glaciology and geophysics to ecology, human health, social sciences, and the humanities; its multilayered organization and planning; and its engagement of new constituencies in the science process—educators, early career scientists, polar residents, and the general public. The committee concludes that the *Vision for the International Polar Year 2007-2008* (NRC, 2004) was realized (Box 1.2 and Box 6.1).

Once established as a concept and plan, IPY developed through the grassroots efforts of researchers, local observers, educators, students, and support personnel from more than 60 nations, including 37 national IPY committees (Krupnik et al., 2011). Contemporary change and compelling science enabled scientists from many disciplines to envision their involvement in IPY (Albert, 2004). An estimated 50,000 researchers, local observers, educators, students, and support personnel were involved in the 228 international IPY projects and in numerous related national efforts (Krupnik et al., 2011). In addition to collaborating internationally, individual scientists and nations were able to focus on their priority issues through their national peer-review funding processes, which also ensured cutting-edge science.

It is significant that IPY was championed early on by two important international organizations, the nongovernmental International Council for Science (ICSU) and the governmental World Meteorological Organization (WMO). Their support served as an international and cross-disciplinary endorsement.

The committee also notes that the U.S. polar research community was well positioned to play a key role in IPY owing to its expertise, resources, and disciplinary breadth. A history of investments in an international focus for U.S. polar research, combined with logistical and scientific strengths, paid off during the planning and preparation for IPY. In addition, strong international professional relationships, spanning many disciplines, enabled U.S. scientists to seize opportunities and take actions to realize goals beyond the capabilities of any single nation.

Examples of successful international collaborations are numerous. Longstanding international colleagues in the ice coring community established the International Partnerships in Ice Core Sciences. New international partnerships, under such banners as the joint U.S.-European Union SEARCH for DAMOCLES (Developing Arctic Modeling and Observing Capabilities for Long-Term Environmental Studies) project and the Arctic Ocean Sciences Board World Climate Research Program's Integrated Arctic Ocean

**BOX 6.1**
**Original IPY Scope and Objectives Statement from *Vision for the International Polar Year 2007-2008***

At its most fundamental level, IPY 2007-2008 is envisioned as an intense, coordinated field campaign of polar observations, research, and analysis that will be multidisciplinary in scope and international in participation. IPY 2007-2008 will be a framework and impetus to undertake projects that could not normally be achieved by any single nation. It allows us to think beyond traditional borders—whether national borders or disciplinary constraints—toward a new level of integrated, cooperative science. A coordinated international approach maximizes both impact and cost-effectiveness, and the international collaborations started today will build relationships and understanding that will bring long-term benefits. Within this context, IPY will seek to galvanize new and innovative observations and research while at the same time building on and enhancing existing relevant initiatives. IPY will serve as a mechanism to attract and develop a new generation of scientists and engineers with the versatility to tackle complex global issues. In addition, IPY is clearly an opportunity to organize an exciting range of education and outreach activities designed to excite and engage the public, with a presence in classrooms around the world and in the media in varied and innovative formats.

The IPY will use today's powerful research tools to better understand the key roles of the polar regions in global processes. Automatic observatories, satellite-based remote sensing, autonomous vehicles, Internet, and genomics are just a few of the innovative approaches for studying previously inaccessible realms. IPY 2007-2008 will be fundamentally broader than past international years because it will explicitly incorporate multidisciplinary and interdisciplinary studies, including biological, ecological, and social science elements .

SOURCE: NRC, 2004.

Observing System (iAOOS), set up shared observing networks throughout the polar seas. The Arctic Observing Network (AON) and the 24-nation Polar Observing Network (POLENET) launched new terrestrial observational networks. Young scientists from a range of disciplines and countries exploited social networking facilities on the Internet to form the Association of Polar Early Career Scientists (APECS) for sharing knowledge and experience. IPY thus provided the impetus for a novel means of collaboration and action among both established and "next-generation" polar scientists and engineers.

## "CHANGE"—THE IPY STRATEGIC MESSAGE

IPY was largely about change: climate system change due to humans, changes in understanding of the polar regions, corresponding changes in research focus, and changes in who does science, how it is done, and how it is communicated. During this time, it became more widely acknowledged that humans are influencing the planet and its climate system, that some changes are occurring faster than anticipated, and that there is a need to take action in response to these changes (i.e., NRC, 2011a).

Historic and current evidence collected during IPY by international teams helped to clarify the impact of human activities in the polar regions. IPY studies yielded important findings about, for example, the continuing dramatic sea ice decline in the Arctic and in the Bellingshausen Sea in the Antarctic; rapid losses of ice in the Greenland ice sheet, on the Antarctic Peninsula, and in coastal areas of West Antarctica; thawing permafrost, terrestrial greening, and biome range changes; and the impacts of climatic warming on ocean circulation and productivity. New sampling also revealed evidence of pollution in remote areas of Antarctica previously thought to be pristine. These and other discoveries during IPY directed scientific inquiry to questions of societal impact, longer-term environmental issues, and sustainability.

In terms of changes in who does science, IPY increased diversity among those involved in the study of the poles. The research community expanded to include more female lead investigators, energetic young scientists launched their own network with the creation of APECS, and Arctic residents and indigenous people's organizations became active participants in the systematic collection of observations.

Methods of research changed as new tools and observational networks supported by new international partnerships increased the ability to detect and document the polar environment. The exploitation of cutting-edge technology and logistics changed understanding of the polar regions by enabling the imaging of previously inaccessible locations across a huge range of spatial scales, from tiny bubbles in thousand-year-old ice to entire mountain ranges under ice sheets. The resulting advances in knowledge of ice sheet formation and flow have profound implications for the ability to predict their future behavior, including their critical contribution to global sea level rise, which has the potential to impact societies all over the world. In addition, a joint project carried out by the United States and United Kingdom (with support from China, Germany, Australia, Japan, and Canada, among others) revealed an alpine environment long hidden beneath the core of the East Antarctic ice sheet, yielding insights into how the ice sheet formed. And the discovery of areas where liquid water beneath the ice sheet freezes onto its underside, providing a significant mechanism for ice sheet growth, was a total surprise.

Evidence of change recorded by international teams of IPY scientists and their local collaborators in polar communities provided vivid content for science education and outreach. IPY science disseminated by new electronic media, special outreach programs, and live communication captured the attention of the public. Dramatic video footage illustrating effects of the changing climate (e.g., Extreme Ice Survey[1] and Polar-Palooza[2]) raised public awareness that what happens at the poles matters to everyone. Participation of U.S. teachers in field work—through programs such as PolarTREC,[3] involvement in the National Science Teachers Association's IPY activities, and international educational linkages through the IPY Programme Office—raised the bar for communication of science to the public and in schools around the world, reaching thousands of schoolchildren. In addition, new standards were established recognizing the responsibility of scientists to communicate to the public and local stakeholders and providing "best practice" methods for doing so.

The Committee has summarized a number of the changes in the perceptions of the polar regions and of polar research that occurred during and because of IPY—see Table 6.1.

## LESSONS LEARNED

With the perspective of several years since the official IPY end date, this committee identified lessons that might inform the planning and organization of future polar research; those lessons are numerous, even as the research results from IPY continue developing.

"International years" are very complex programs. Past such endeavors (e.g., the International Year of Physics in 2005, International Heliophysical Year during 2007 through 2009, International Year of Astronomy in 2009, International Year of Biodiversity in 2010, and International Year of Forests in 2011) have covered a wide variety of important topics. To take flight from enthusiasts' dreams and drawing boards, each "international year" needs energetic, well-placed and -connected individuals and teams to bring together the multiple entities that will become the essential components of the global venture.

For IPY 2007-2008, the concept of an "international year" proved to be as valid for today's highly dynamic global science as it was 50, 75, and 125 years ago. International years give a higher level of visibility, allow greater breadth of work and implementation of infrastructure, and increase the leverage and "esprit de corps" of the science community. These attributes helped to make this IPY an exciting once-in-a-lifetime event, not just for researchers, but also for students, journalists, and members of the general public.

The inclusiveness of the IPY planning process and implementation was a strong motivator for broad participation and a powerful driver of the IPY success. Science initiatives with a specific focus sometimes are (or appear to be) exclusive—to other disciplines, to nonscientists, or to nonparticipating nations. The explicitly inclusive approach adopted by the IPY planners and coordinators unleashed the energy of volunteerism, new partnerships, and cross-boundary communication. It helped bring down barriers between science fields, between scientists and polar residents, between professional researchers and science educators, and between nations with and without significant previous engagement in polar research. Not all future initiatives may achieve the breadth of IPY 2007-2008, but those that strive to be inclusive will have a greater and longer-lasting impact.

This IPY revealed the level of effort required of the core enthusiasts to convince and engage the community at large. It illustrated the importance of careful planning, inclusiveness, effective mobilization of the energy and ideas of hundreds of volunteers, and good timing. It also underscored the amount of time needed

[1] *www.extremeicesurvey.org.*
[2] *http://passporttoknowledge.com/polar-palooza.*
[3] *www.polartrec.com.*

TABLE 6.1 Changing Perceptions of the Polar Regions and of Polar Research

| **Before IPY** | **During and After IPY** |
|---|---|
| *White*—sea ice, glaciers, snow-covered tundra | *Dark*—with the open water in the Arctic as sea ice retreats, barren land revealed by melting valley glaciers, lakes on top of glaciers, northward migration of shrubs and boreal forests |
| Frigid/cold | *Warming*—as shown by contours of recent and projected future terrestrial and marine warming |
| Frozen/icy | *Melting/thawing*—as vividly seen in the glacial moulins, retreating ice-cored coasts, and structures undermined by thawing permafrost |
| *Static/slow*—ice "cap," "perma" frost, compact Arctic vortex, stable fisheries | *Dynamic/fast*—old sea ice blown out of the Arctic, Greenland and Antarctic ice streams accelerating, icebergs calving, lobes of winter Arctic air penetrating south, advancing/replacement fisheries |
| Pristine | *Contaminated*—ozone hole, evidence that cold trapping results in high concentrations of organochlorides in polar bears, evidence of industrial airborne pollution in the Greenland ice sheet, Asian and North American sources of particulates in troposphere |
| *Robust*, intimidating, stable, thick ice | *Vulnerable* |
| *Inaccessible*—thick sea ice hampers access to the interior | *Accessible* for a fee—tourism along previously ice-clogged coasts of Greenland, the Canadian Archipelago, and the Antarctic Peninsula; airplane access to the North and South Poles |
| *Domain of the residents*—indigenous peoples in the north, scientists in the south | *Collaborative networks*—increased online partnership and data sharing |
| *Remote* and disconnected | *Connected*—global sea level rise from glacier melt, "warm Arctic/cold continent" leading to weather changes in highly populated temperate zones |
| *Peripheral*—literally "off the map" | *Central*—warming first and fastest, 2-3 times amplification |
| *Disciplinary*, multidisciplinary | Increasingly *interdisciplinary*—complex in terms of systems, participatory |
| Basic research | Research increasingly driven by *applications* |
| Spatial and temporal *distributions* and variability | *Trends, thresholds, tipping points, global feedback*, studying change while experiencing change |
| *Arctic and Antarctic as separate*, opposing domains ("polar bears" and "penguins") | *Bipolar science*, growing exchange, Arctic-Antarctic connections and partnerships |
| Expeditions and multiyear initiatives | In situ long-term *observations and monitoring* by local residents |
| *Models* seen as worst case | *Reality is worse* than models (heading above IPCC's "A1B" scenario, sea ice loss, glacier mass loss) |
| Scientists primarily *male* | Increasing number of *female* participants and leaders |
| Established researchers | Energized *next generation* of researchers, newly formed Association of Polar Early Career Scientists |
| *Education and outreach* a duty, an add-on | *Education and outreach integrated* with and feeding back into research |
| Public *perception* of poles as remote or "cute" | Public *awareness* of changes, interest and concern |
| Established specialist community with tendency to *national* focus | *International* focus with many new players |

SOURCE: Compiled by the Committee based on Committee members' experience as researchers and educators and on interactions with the public.

for preparation (4-5 years before the actual research and observation period), the value of timely endorsement by and involvement of a broad spectrum of the science community, and the role of respected international institutional leadership—in this case ICSU and WMO—to ensure proper governance and visibility.

A key organizational lesson of IPY is the critical role of seed resources for planning and implementation. The original investment by ICSU in the IPY planning in early 2003 was $60,000—a fraction of a percent of the total estimated funds used in IPY as a whole. That investment supported the ICSU Planning Group of 14 international members who produced the crucial overarching IPY framework from July 2003 to October 2004.

In addition to funds from ICSU and WMO, the U.S. contribution was critical to early planning, with the timely injection of $200,000 in the spring of 2003 by the National Academy of Sciences (NAS) to support the U.S. National IPY Committee. This support resulted in the National Research Council (NRC) document, *A Vision for the International Polar Year 2007-2008* (NRC, 2004), that was instrumental in mobilizing the U.S. science community and agencies for IPY. NSF, for example, referenced the NRC report multiple times in its calls for proposals, and advised applicants that "IPY proposals are expected to help implement the vision developed by and articulated in the [Vision Report]."[4]

The NRC also held an implementation workshop for federal agency representatives, members of the NRC Polar Research Board (PRB), and members of the U.S. National Committee for IPY to talk about how the United States might address the scientific challenges articulated in the Vision Report (NRC, 2004) and how to move ahead in developing a suite of coordinated scientific activities (NRC, 2005).

Support from NSF, National Oceanic and Atmospheric Administration (NOAA), and Cooperative Institute for Research in Environmental Sciences (CIRES[5]) for education and outreach brainstorming workshops in 2004 and 2005 galvanized and organized the polar community, helping them prepare for an extensive and effective educational campaign to accompany the IPY science programs and findings.

U.S. federal agency involvement in IPY was led by NSF, which committed over $347 million[6] for science and education activities, including a $60 million appropriation from Congress. The National Aeronautics and Space Administration, NOAA, and the U.S. Geological Survey also funded IPY-related programs, many of which had significant international partnerships. In addition, a plethora of smaller but innovative programs arose from endowed university-based programs, museums, and other nonprofit organizations; these contributed to IPY outreach and increased public engagement.

In 2005, additional ICSU and WMO investments of $250,000 supported the IPY Joint Committee of 20 members, who steered the science preparation, implementation, and completion of IPY in 2005-2010. The daily tasks of managing the international IPY activities via the International Programme Office (IPO) in Cambridge, UK, were supported by a UK national contribution of approximately $1.5 million (again, a small percentage of the estimated total funding for IPY as a whole). These modest investments leveraged additional national funding from all countries for research in 2006-2009 that totaled more than $1.2 billion for IPY as a whole.[7]

As might be expected with any large-scale, complex endeavor, some challenges and difficulties arose. For example, despite valiant attempts by the IPY Data Committee and several coordinating workshops, the development and accessibility of IPY data products were hampered by a shortage of time and resources. As a result, this committee relied as best they could on international coordination and negotiation using existing data systems and management structures.

More effective interagency coordination within and across nations, particularly in funding approval and logistics, would have been beneficial. Not all scientific research priorities received adequate support (anecdotally, climate modeling has been mentioned as one such area), in part because of inherent difficulties in coordinating research from the top down, whereas decisions about which projects will be funded often come from a merit-based (i.e., more bottom-up) system. Delays in national funding processes affected abilities to coordinate field research and infrastructure sharing. A formal mechanism for interaction of representatives of funding agencies from many nations and the international community-based planning committee(s) would have been helpful. In the end, leads of federal funding agencies from the United States and other nations forged agreements through their own initiative to enable the success of large international programs.

The lack of continued support to coordinate IPY-initiated programs has made it difficult to maintain the full scope of valuable researcher, funding, and innovation networks developed and nurtured during IPY. Useful components of the larger IPY structure—such as the international IPY website, its publication database, and educational/outreach efforts—have struggled to find alternative resources, and funding could have maintained the U.S. IPY website as a more consistently useful resource during IPY. Overall, the sustained impact and momentum of the IPY legacy will require

[4] Section IX, p. 22, of Program Solicitation NSF-06-534.

[5] *http://cires.colorado.edu/*.

[6] This estimate includes awards made over four fiscal years 2006-2009, with more than half the money going out during the 2007-2009 IPY field period. Many awards, though, were of several years' duration—to accommodate laboratory work and other follow-up after return from the field—so final funding increments for some of the later ones will not be sent until 2013.

[7] Estimates of total IPY funding vary from approximately $1.2 billion (not counting many national polar infrastructure investments; Krupnik et al., 2011) to approximately $1.5 billion (Carthage Smith, International Council for Science, personal communication, 2011).

ongoing support from funding agencies for both the observing networks and the scientists.

The Vision Report defined the terms for U.S. IPY efforts, and this follow-up report on lessons and legacies provides the concluding bookend for those efforts. Several other nations (Canada, Sweden, Japan, the United Kingdom, Norway) invested in the production of timely assessments of their national IPY activities similar to this report, whereas others closed their IPY programs without a concluding assessment or statement.

The authors of this report urge planners of the next IPY (or of similar international efforts) to consider all of the lessons identified in this report, as well as the mix of ingredients that made IPY such a success.

## LEGACIES

IPY changed perceptions and understanding of the polar regions. Its findings revealed that the Earth system cannot be understood without knowledge of the dynamics of these regions, a message that is especially relevant in light of evidence of the many global impacts of polar change. It also became clear during IPY that traditional knowledge can make a material contribution to the joint assessment of global processes and that science and scientists can provide effective means of achieving international discourse. At a time when the polar regions, in particular the Arctic, are undergoing a transformation from a perceived icy wilderness to a new zone for human affairs, these new insights could not be more timely or relevant.

The success of IPY was also evident in the people it touched. The international polar research community grew in terms of inclusiveness, capability, and experience. Arctic residents, and particularly indigenous communities, learned that information from science and scientists can be used to inform and enrich their daily lives. For their part, scientists learned how to make the results of their science useful for decisions faced by citizens of both the Arctic and the midlatitudes. Students and public audiences in numerous countries became engaged in learning about the current climate change that affects all people, and in the thrill and excitement of unraveling the mysteries of the planet and its extraordinary polar environments.

For all these reasons, IPY was a success scientifically, organizationally, and as a collective international endeavor as humanity grapples with the complexities and challenges of the many changes occurring in the environment and societies around the world. May it provide an inspiration for planners of the future, as science increasingly provides the knowledge that informs action.

# References

ACCAP (Alaska Center for Climate Assessment and Policy). 2010. Decision-Making for At-Risk Communities in a Changing Climate. Fairbanks: Alaska Center for Climate Assessment and Policy, University of Alaska.

ACIA (Arctic Climate Impact Assessment). 2005. Arctic Climate Impact Assessment: Scientific Report. Cambridge, UK, and New York: Cambridge University Press.

AHDR (Arctic Human Development Report) 2004. Arctic Human Development Report. Akureyri, Iceland: Stefansson Arctic Institute.

Aksnes, D. W., and D. O. Hessen. 2009. The structure and development of polar research (1981-2007): A publication-based approach. Arctic, Antarctic, and Alpine Research 41(2):155-163.

Albert, M. 2004. The International Polar Year. Science 303(5663):1437.

Albert, M., E. Bell, Z. Courville, J. McConnell, K. Mueller, S. Tronstad, H. Anschuetz, T. Neumann, and T. Scambos. 2012. Polar firn properties reflect accumulation rate across Antarctica. Journal of Geophysical Research (in review).

Allen, J., G. V. Mohatt, C. C. Fok, D. Henry, and People Awakening Team. 2009. Suicide prevention as a community development process: Understanding circumpolar youth suicide prevention through community level outcomes. International Journal of Circumpolar Health 68(3):274-291.

Alley, R. B., J. T. Andrews, J. Brigham-Grette, G. K. C. Clarke, K. M. Cuffey, J. J. Fitzpatrick, S. Funder, S. J. Marshall, G. H. Miller, J. X. Mitrovica, D. R. Muhs, B. L. Otto-Bliesner, L. Polyak, and W. C. White. 2010. History of the Greenland ice sheet: Paleoclimatic insights. Quaternary Science Reviews 29(15-16):1728-1756.

AMAP (Arctic Monitoring and Assessment Program). 2009. Arctic Monitoring and Assessment Program Report. Arctic Pollution 2009. Oslo, Norway: Arctic Monitoring and Assessment Program.

Andersen, T., J. Kruse, and B. Poppel. 2002. Survey of the Living Conditions in the Arctic: Inuit, Saami and the Indigenous Peoples of Chukotka (SLICA). Arctic 55(3):310-317.

Anschutz, H., K. Muller, E. Isaksson, J. R. McConnell, H. Fischer, H. Miller, M. Albert, and J.-G. Winther. 2009. Revisiting site of the South Pole Queen Maud Land Traverses in East Antarctica: Accumulation data from shallow firn cores. Journal of Geophysical Research 114:D24106.

Aporta, C. 2009. The trail as home: Inuit and their pan-Arctic network of routes. Human Ecology 37(2):131-146.

Arblaster, J. M., G. A. Meehl, and D. J. Karoly. 2011. Future climate change in the Southern Hemisphere: Competing effects of ozone and greenhouse gases. Geophysical Research Letters 38:L02701.

Arctic Council. 2009. Arctic Council Arctic Marine Shipping Assessment 2009 Report. Tromsø, Norway: Arctic Council.

Arrigo, K. R., G. V. Dijken, and S. Pabi. 2008. Impact of a shrinking Arctic ice cover on marine primary production. Geophysical Research Letters 35:L19603.

Avango, D., L. Hacquebord, Y. Aalders, H. D. Haas, U. Gustafsson, and F. Kruse. 2011. Between markets and geo-politics: natural resource exploitation on Spitsbergen (Svalbard), from 1600 to the present day. Polar Record 47(240):29-39.

Barber, D. G., and D. Barber. 2007. Two ways of knowing: Merging science and traditional knowledge during the Fourth International Polar Year. Winnepeg: University of Manitoba.

Barr, S. and P. Chaplin, eds. 2008. Historical Polar Bases—Preservation and Management. Monuments and Sites XVIII. Paris, France: International Polar Heritage Committee of the International Council on Monuments and Sites.

Barr, S., and C. Luedecke, eds. 2010. The History of the International Polar Years (IPYs). Dordrecht, The Netherlands: Springer.

Bates, N., and J. T. Mathis. 2009. The Arctic Ocean marine carbon cycle: Evaluation of air-sea $CO_2$ exchanges, ocean acidification impacts and potential feedbacks. Biogeosciences 6(11):2433-2459.

Bates, N., W. J. Cai, and J. Mathis. 2011. The ocean carbon cycle in the western Arctic Ocean: Distributions and air-sea fluxes of carbon dioxide. Oceanography 24(3):186-201.

Bell, R., F. Ferraccioli, T. T. Creyts, D. Braaten, H. Corr, I. Das, D. Damaske, N. Frearson, T. Jordan, K. Rose, M. Studinger, and M. Wolovick. 2011. Widespread persistent thickening of the East Antarctic ice sheet by freezing from the base. Science 331(6024):1592-1595.

Berger, A., and M. F. Loutre. 1991. Insolation values for the climate of the last 10 million years. Quaternary Science Reviews 10(4):297-317.

Berkman, P. A. 1960. President Eisenhower, the Antarctic Treaty and the origin of international spaces. In Science Diplomacy: Antarctica, Science and the Governance of International Spaces, P. A. Berkman, M. A. Lang, D. W. H. Walton, and O. R. Young, eds. Washington, DC: Smithsonian Institution Scholarly Press.

Berkman, P. A., M. A. Lang, D. W. H. Walton and O. R. Young, Eds. 2011. Science Diplomacy: Antarctica, Science and the Governance of International Spaces. Washington, DC: Smithsonian Institution Scholarly Press.

Bhatt, U. S., D. A. Walker, M. K. Raynolds, J. C. Comiso, H. E. Epstein, G. S. Jia, R. Gens, J. E. Pinzon, C. J. Tucker, C. E. Tweedie, and P. J. Webber. 2010. Circumpolar Arctic tundra vegetation change is linked to sea ice decline. Earth Interactions 14. available at *http://www.geobotany.org/library/pubs/BhattUS2010_ei_1-20.pdf.*

Bindschadler, R. A., P. Vornberger, A. Fleming, A. Fox, J. Mullins, D. Binnie, S. J. Paulsen, B. Granneman, and D. Gorodetzky. 2008. The Landsat Image Mosaic of Antarctica. Remote Sensing Environment 112(12):4214-4226.

Bindschadler, R. A., D. G. Vaughan, and P. L. Vornberger. 2011. Variability of basal melt beneath the Pine Island Glacier ice shelf, West Antarctica. Journal of Glaciology 57(204):581-595.

Bird, K. J., R. Charpentier, D. Gautier, D. Houseknecht, T. Klett, J. Pitman, T. Moore, C. Schenk, M. Tennyson, and C. Wandrey. 2008. Circum-Arctic resource appraisal: Estimates of undiscovered oil and gas north of the Arctic Circle. Denver: U.S. Geological Survey Information Services.

Bisiaux, M. M., R. Edwards, J. R. McConnell, M. R. Albert, H. Anschütz, T. A. Neumann, E. Isaksson, and J. E. Penner. 2012. Variability of black carbon deposition to the East Antarctic Plateau, 1800-2000 AD. Atmospheric Chemistry and Physics 12:3799-3808.

Biuw, M., L. Boehme, C. Guinet, M. Hindell, D. Costa, J. B. Charrassin, F. Roquet, F. Bailleul, M. Meredith, S. Thorpe, Y. Tremblay, B. McDonald, Y.-H. Park, S. R. Rintoul, N. Bindoff, M. Goebel, D. Crocker, P. Lovell, J. Nicholson, F. Monks, and M. A. Fedak. 2007. Variations in behavior and condition of a Southern Ocean top predator in relation to *in situ* oceanographic conditions. Proceedings of the National Academy of Sciences of the United States of America 104(34):13705-13710.

Bluhm, B. A., K. Iken, S. M. Hardy, B. I. Sirenko, and B. A. Holladay. 2009. Community structure of epibenthic megafauna in the Chukchi Sea. Aquatic Biology 7(3):269-293.

Bluhm, B. A., A. V. Gebruk, R. Gradinger, R. R. Hopcroft, F. Huettmann, K. N. Kosobokova, B. I. Sirenko, and J. M. Weslawski. 2011. Arctic marine biodiversity: An update of species richness and examples of biodiversity change. Oceanography 24(3):232-248.

Børresen, M. L., O. R. Olsen, K. Ladefoged, B. J. McMahon, H. Krarup, and A. Koch. 2010. Hepatitis D outbreak among children in a hepatitis B hyper-endemic settlement in Greenland. Journal of Viral Hepatitis 17(3):162-170.

Bravo, M. T. 2010. Epilogue: The humanism of sea ice. In SIKU: Knowing Our Ice, I. Krupnik, C. Aporta, S. Gearheard, G. J. Laidler, and L. Kielsen Holm, eds. Dordrecht, The Netherlands: Springer.

Brigham-Grette, J., M. Melles, P. Minyuk, and C. Koeberl. 2011. Millennial-scale Arctic climate change of the last 3.6 million years: Scientific drilling at Lake El'gygytgyn, northeast Russia. Oceanography 24(3):80-81.

Broadbent, N. 2009. From Ballooning in the Arctic to 10,000-Foot Runways in Antarctic: Lessons from Historic Archaeology. In Smithsonian at the Poles: Contributions to International Polar Year Science. I. Krupnik, M. A. Lang, and S. E. Miller, eds. Washington, D.C.: Smithsonian Institution.

Brock, C. A., J. Cozic, R. Bahreini, K. D. Froyd, A. M. Middlebrook, A. McComiskey, J. Brioude, O. R. Cooper, A. Stohl, K. C. Aikin, J. A. de Gouw, D. W. Fahey, R. A. Ferrare, R.-S. Gao, W. Gore, J. S. Holloway, G. Hübler, A. Jefferson, D. A. Lack, S. Lance, R. H. Moore, D. M. Murphy, A. Nenes, P. C. Novelli, J. B. Nowak, J. A. Ogren, J. Peischl, R. B. Pierce, P. Pilewskie, P. K. Quinn, T. B. Ryerson, K. S. Schmidt, J. P. Schwarz, H. Sodemann, J. R. Spackman, H. Stark, D. S. Thomson, T. Thornberry, P. Veres, L. A. Watts, C. Warneke, and A. G. Wollny. 2011. Characteristics, sources, and transport of aerosols measured in spring 2008 during the Aerosol, Radiation, and Cloud Processes affecting Arctic Climate (ARCPAC) Project. Atmospheric Chemistry and Physics 11:2423-2453.

Bromwich, D. H., and J. P. Nicolas. 2010. Sea-level rise: Ice-sheet uncertainty. Nature Geoscience 3(9):596-597.

Bromwich, D. H., Y.-H. Kuo, M. Serreze, J. Walsh, L.-S. Bai, M. Barlage, K. M. Hines, and A. Slater. 2010. Arctic system reanalysis: Call for community involvement. Eos Transactions, American Geophysical Union 92(2):13-14.

Brooks, S., C. Moore, D. Lew, B. Lefer, G. Huey, and D. Tanner. 2011. Temperature and sunlight controls of mercury oxidation and deposition atop the Greenland ice sheet. Atmospheric Chemistry and Physics 11:3663-3691.

Brown, J., and V. E. Romanovsky. 2008. Report from the International Permafrost Association: State of permafrost in the first decade of the 21st century. Permafrost and Periglacial Processes 19(2):255-260.

Bruce, M. G., S. L. Deeks, T. Zulz, D. Bruden, C. Navarro, M. Lovegren, L. Jette, K. Kristinsson, G. Sigmundsdottir, K. B. Jensen, O. Lovoll, J. P. Nuorti, E. Herva, A. Nystedt, A. Sjostedt, A. Koch, T. W. Hennessey, and A. J. Parkinson. 2008. International circumpolar surveillance for invasive pneumococcal disease, 1999-2005. Emerging Infectious Diseases 14(1):25-33.

Caldeira, K., and P.B. Duffy. 2000. The role of the Southern Ocean in uptake and storage of anthropogenic carbon dioxide. Science 287:620-622.

Calder, J., H. Eicken, and J. Overland. 2011. The sea outlook. In Understanding Earth's Polar Challenges: International Polar Year 2007-2008, I. Krupnik, ed. Edmonton, Canada: CCI Press.

Calef, M. P. 2010. Recent climate change impacts on the boreal forest of Alaska. Geography Compass 4(2):67-80.

Cao, L., and K. Caldeira. 2008. Atmospheric $CO_2$ stabilization and ocean acidification. Geophysical Research Letters 35:L19609.

Carlson, D. 2011. A lesson in sharing. Nature 469(7330):293.

Carnegie Foundation. 2008. The Arctic Climate Change and Security Policy Conference: Final Report and Findings. December 1-3, 2008, Dartmouth College, Hanover, New Hampshire. Washington, DC: Carnegie Endowment for Internaional Peace.

Cavalieri, D. J., and C. L. Parkinson. 2008. Antarctic sea ice variability and trends, 1976-2006. Journal of Geophysical Research 113:C07004.

CCSP (U.S. Climate Change Science Program). 2009. Synthesis and Assessment Product 1.2: Past Climate Variability and Change in the Arctic and at High Latitudes. Washington, DC: U.S. Geological Survey.

Chan, L. H. M., G. M. Egeland, I. Sobol, K. Young, P. Ayotte, and G. Stern. 2009. Assessment of contaminant and dietary nutrient interactions in Inuit Health Survey: Nunavut, Nunatsiavut and Inuvialuit. Pp. 8-13 in Synopsis of Research Conducted Under the 2008-2009 Northern Contaminants Program. Ottawa, Canada: Department of Indian Affairs and Northern Development.

Chao, B. F., Y. Wu, and Y. S. Li. 2008. Impact of artificial reservoir water impoundment on global sea level. Science 320(5873):212-214.

Chapin, F. S., and A. L. Lovecraft. 2011. Contextualizing Alaska's climate change from global to local scales: The boreal forest, people and wildfire. In North by 2020: Perspectives on Alaska's Changing Social-Ecological Systems, A. L. Lovecraft and H. Eicken, eds. Fairbanks: University of Alaska Press.

Chapin, F. S., A. D. McGuire, R. W. Ruess, T. N. Hollingsworth, E. S. Euskirchen, J. F. Johnstone, J. B. Jones, M. T. Jorgenson, E. S. Kasischke, K. Kielland, G. P. Kofinas, A. H. Lloyd, M. C. Mack, D. L. Taylor, M. R. Turetsky, and J. Yarie. 2010. Resilience and change in Alaska's boreal forest. Canadian Journal of Forest Research 40:1336-1346.

Christiansen, H. H., B. Etzelmuller, K. Isaksen, H. Juliussen, H. Farbrot, O. Humlum, M. Johansson, T. Ingeman-Nielsen, L. Kristensen, J. Hjort, P. Holmlund, A. B. K. Sannel, C. Sigsgaard, H. J. Akerman, N. Foged, L. H. Blikra, M. A. Pernosky, and R. S. Odegard. 2010. The thermal state of permafrost in the nordic area during the International Polar Year 2007-2009. Permafrost and Periglacial Processes 21(2):156-181.

Church, J. A., and N. J. White. 2006. A 20th-century acceleration in global sea-level rise. Geophysical Research Letters 33:L01602.

CIRES and NOAA (Cooperative Institute for Research in Environmental Sciences and National Oceanic and Atmospheric Administration). 2005. Poles Together: Coordinating International Polar Year (IPY) Outreach and Education, Summary Report. Available at *http://cires.colorado.edu/education/outreach/ipyoe/PolesTogetherSummary.pdf*.

Clinton, H. 2009. Secretary Clinton on the Antarctic Treaty. Available at *http://www.realclearpolitics.com/articles/2009/04/secretary_clinton_on_the_antar.html*.

Coastal Response Research Center. 2009. Opening the Arctic Seas: Envisioning Disasters and Framing Solutions. Durham: University of New Hampshire.

Costa, D., J. Klinck, E. Hoffman, M. Dinniman, and J. Burns. 2008. Upper ocean variability in west Antarctic Peninsula continental shelf waters as measured using instrumented seals. Deep Sea Research Part II: Topical Studies in Oceanography 55(3-4):323-337.

Crate, S. A., B. C. Forbes, L. King, and J. Kruse. 2010. Contact with nature. Pp. 109-127 in Arctic Social Indicators: A Follow-up to the Arctic Human Development Report, J. N. Larsen, P. Schweitzer, and G. Fondahl, eds. Copenhagen: Nordic Council of Ministers.

Das, S., I. Joughin, M. Behn, I. Howatt, M. King, D. Lizarralde, and M. Bhatia. 2008. Fracture Propagation to the base of the Greenland ice sheet during supraglacial lake drainage. Science 320(5877):778.

Denton, G. H., D. E. Sugden, D. R. Marchant, B. L. Hall, and T. I. Wilch. 1993. East Antarctic ice sheet sensitivity to Pliocene climatic change from a Dry Valleys perspective. Geografiska Annaler: Series A, Physical Geography 75(4):155-204.

Dewailly, É., G. Muckle, P. Ayotte, B. Valera, and M. Noël. 2009. Contaminant exposure and emergence of cardio-vascular diseases in Inuit. Pp. 24-41 in Synopsis of Research Conducted Under the 2008-2009 Northern Contaminants Program, S. Smith, J. Stow, and J. Edwards, eds. Ottawa, Canada: Department of Indian Affairs and Northern Development.

Dibb, J. E., L. D. Ziemba, J. Luxford and P. Beckman. 2010. Bromide and other ions in the snow, firn air, and atmospheric boundary layer at Summit during GSHOX. Atmospheric Chemistry and Physics 10:9931-9942.

Dickson, B., and E. Fahrbach. 2011. Observing our northern seas during the IPY: What was achieved, what have we learned, where do we go from here? In Understanding Earth's Polar Challenges: International Polar Year 2007-2008, I. Krupnik, I. Allison, R. Bell, P. Cutler, D. Hik, J. López-Martínez, V. Rachold, E. Sarukhanian, and C. Summerhayes, eds. Edmonton, Canada: CCI Press.

Dickson, B., B. Rudels, C. Lee, T. Haine, and iAOOS PIs. 2010. iAOOS: An Ocean-Observing System for Northern Seas During the Legacy Phase of the International Polar Year: A Report of the Arctic Ocean Sciences Board. Bremerhaven, Germany: Alfred Wegener Institute for Polar and Marine Research.

Ding, Q., E. J. Steig, D. Battisti, and M. Küttel. 2011. Winter warming in West Antarctica caused by central tropical Pacific warming. Nature Geoscience 4:398-403.

Dinniman, M., J. Klinck, and W. Smith, Jr. 2011. A model study of Circumpolar Deep Water on the West Antarctic Peninsula and Ross Sea continental shelves. Deep Sea Research Part II: Topical Studies in Oceanography 58(13-16):1508-1523.

Dmitrenko, I. A., I. V. Polyakov, S. A. Kirillov, L. A. Timokhov, I. E. Frolov, V. T. Sokolov, H. L. Simmons, V. V. Ivanov, and D. Walsh. 2008. Toward a warmer Arctic Ocean: Spreading of the early 21st century Atlantic Water warm anomaly along the Eurasian Basin margins. Journal of Geophysical Research 113:C05023.

Dmitrenko, I. A., D. Bauch, S. A. Kirillov, N. Koldunov, P. J. Minnett, V. V. Ivanov, J. A. Hölemann, and L. A. Timokhov. 2009. Barents Sea upstream events impact the properties of Atlantic Water inflow into the Arctic Ocean: Evidence from 2005 to 2006 downstream observations. Deep Sea Research Part I: Oceanographic Research Papers 56(4):513-527.

Doherty, S. J., S. G. Warren, T. C. Grenfell, A. D. Clarke, and R. E. Brandt. 2010. Light-absorbing impurities in Arctic snow. Atmospheric Chemistry and Physics 10:11647-11680.

Domack, E. W., A. Leventer, S. Root, J. Ring, E. Williams, D. Carlson, E. Hirshorn, W. Wright, R. Gilbert, and G. Burr. 2003. Marine sedimentary record of natural environmental variability and recent warming in the Antarctic Peninsula. In Antarctic Peninsula Climate Variability: Historical and Paleoenvironmental Perspectives, E. Domack, A. Leventer, B. Adam, R. Bindschadler, P. Convey, and M. Kirby, eds. Washington, DC: American Geophysical Union.

Doney, S. C. 2006. The dangers of ocean acidification. Scientific American 294(3):58-65.

Dorrepaal, E., S. Toet, R. S. P. van Logtestijn, E. Swart, M. J. van de Weg, T. V. Callaghan, and R. Aerts. 2009. Carbon respiration from subsurface peat accelerated by climate warming in the subarctic. Nature 460:616-619.

Downie, D. L., and T. Fenge, eds. 2003. Northern Lights against POPs: Combatting Toxic Threats in the Arctic. Montreal, Quebec and Kingston, Ontario: McGill-Queen's University Press.

Druckenmiller, M. L., H. Eicken, J. C. George, and L. Brower. 2010. Assessing the shorefast ice: Iñupiat whaling trails off Barrow, Alaska. In SIKU: Knowing Our Ice, I. Krupnik, C. Aporta, S. Gearheard, G. J. Laidler, and L. Kielsen Holm, eds. Dordrecht, The Netherlands: Springer.

Durner, G. M., D. Douglas, R. Nielson, S. Amstrup, T. McDonald, I. Stirling, M. Mauritzen, E. Born, Ø. Wiig, E. DeWeaver, M. Serreze, S. Belikov, M. Holland, J. Maslanik, J. Aars, D. Bailey, and A. Derocher. 2009. Predicting 21st-century polar bear habitat distribution from global climate models. Ecological Monographs 79:25-58.

Egeland, G. 2009. Nunatsiavut Inuit Health Survey: "Kanuivit?" Pp. 42-44 in Synopsis of Research Conducted Under the 2008-2009 Northern Contaminants Program, S. Smith, J. Stow, and J. Edwards, eds. Ottawa, Canada: Deparment of Indian Affairs and Northern Development.

Eicken, H. 2010. Indigenous knowledge and sea ice science. In SIKU: Knowing Our Ice, I. Krupnik, C. Aporta, S. Gearheard, G. J. Laidler, and L. Kielsen Holm, eds. Dordrecht, The Netherlands: Springer.

Eicken, H., A. L. Lovecraft, and M. L. Druckenmiller. 2009. Sea-ice system services: A framework to help identify and meet information needs relevant for Arctic observing networks. Arctic 62(2):119-136.

Eicken, H., G. Hufford, V. Metcalf, S. Moore, J. Overland, and H. Wiggins. 2011. Sea Ice for Walrus Outlook (SIWO). In Understanding Earth's polar challenges: International Polar Year 2007-2008, I. Krupnik, Allison, R. Bell, P. Cutler, D. Hik, J. López-Martínez, V. Rachold, E. Sarukhanian, and C. Summerhayes, eds. Edmonton: CCI Press.

Elzinga, A. 2009. Through the lens of the polar years: Changing characteristics of polar research in historical perspective. Polar Record 45(235):313-336.

Farahani, E., N. McFarlaner, R. L. Batchelor, R. L. Collins, V. L. Harvey, N. J. Livesy, G. L. Manney, M. L. Santee, K. Strong, and K. A. Walker. 2009. Features of the Arctic stratosphere during IPY. SPARC Newsletter 33:6-13.

Ferraccioli, F., C. Finn, T. Jordan, R. Bell, L. Anderson, and D. Damaske. 2011. East Antarctic rifting triggers uplift of the Gamburtsev Mountains. Nature 479:388-392.

Fitzpatrick, J., R. Alley, J. Brigham-Grette, G. Miller, L. Polyak, and J. W. C.White. 2010. Arctic paleoclimate synthesis thematic papers introduction. Quaternary Science Reviews 29(15-16):1674-1678.

Forbes, B. C., F. Stammler, T. Kumpula, N. Meschtyb, A. Pajunen, and E. Kaarlejarvi. 2009. High resilience in the Yamal-Nenets social-ecological system, West Siberian Arctic, Russia. Proceedings of the National Academy of Sciences of the United States of America 106(52):22041-22048.

Fowler, C., W. Emery, and J. A. Maslanik. 2003. Satellite-derived Arctic sea ice age: October 1978 to March 2003. Geoscience and Remote Sensing Letters 1(2):71-74.

Francis, J. A., C. Weihan, D. Leathers, J. Miller, and D. Veron. 2009. Winter Northern Hemisphere weather patterns remember summer Arctic sea-ice extent. Geophysical Research Letters 36:L07503.

Fretwell, P., and P. N. Trathan. 2009. Penguins from space: Faecal stains reveal the location of emperor penguin colonies. Global Ecology and Biogeography 18(5):543-552.

Frey, K. E., and J. W. McClelland. 2009. Impacts of permafrost degradation on Arctic river biogeochemistry. Hydrological Processes 23:169-182.

Friedlaender, A., D. Johnston, W. Fraser, J. Burns, P. N. Halpin, and D. Costa. 2011. Ecological niche modeling of sympatric krill predators around Marguerite Bay, western Antarctic Peninsula. Deep Sea Research Part II: Topical Studies in Oceanography 58(13-16):1729-1740.

Funder, S., H. Goosse, H. Jepsen, E. Kaas, K. H. Kjær, N. J. Korsgaard, N. Larsen, H. Linderson, A. Lyså, P. Möller, J. Olsen, and E. Willerslev. 2011. A 10,000 year record of Arctic Ocean sea ice variability—view from the beach. Science 333(6043):747-750.

Galloway, T., T. K. Young, and G. Egeland. 2010. Emerging obesity among preschool-aged Canadian Inuit children. International Journal of Circumpolar Health 69(2):151-157.

GAO (U. S. Government Accountability Office). 2009. Alaska Native Villages: Limited Progress Has Been Made on Relocating Villages Threatened by Flooding and Erosion. Washington, DC: U.S. Government Accountability Office.

Gauthier, M., M. Simard, and B. Blais. 2010. Prevalence of *Escherichia coli* O157:H7 and *Salmonella* in traditional meats derived from game animals in Nunavik. Rural and Remote Health 10(2):1329.

Gearheard, S., C. Aporta, G. Aipellee, and K. O'Keefe. 2011. The Igliniit Project: Inuit hunters document life on the trail to map and monitor Arctic change. The Canadian Geographer 55(1):42-55.

Gesink-Law, D., E. Rink, R. Montgomery-Anderson, G. Mulvad, and A. Koch. 2010. Developing a culturally compotent and sociallyh relevant sexual health survey with an urban Arctic community. International Journal of Circumpolar Health 69(1):25-37.

Gesink-Law, D., E. Rink, G. Mulvad, and A. Koch. 2008. Sexual health and sexually transmitted infections in the North American Arctic. Emerging Infectious Diseases 14(1):4-9.

Gofman, V., and C. Dickson. 2011. Engaging Arctic indigenous organizations in IPY 2007-2008. In Understanding Earth's Polar Challenges: International Polar Year 2007-2008, I. Krupnik, I. Allison, R. Bell, P. Cutler, D. Hik, J. López-Martínez, V. Rachold, E. Sarukhanian, and C. Summerhayes, eds. Edmonton, Canada: CCI Press.

Gradinger, R. 2009. Sea-ice algae: Major contributors to primary production and algal biomass in the Chukchi and Beaufort Seas during May/June 2002. Deep Sea Research Part II: Topical Studies in Oceanography 56(17):1201-1212.

Grebmeier, J. M. 2012. Shifting patterns of life in the Pacific Arctic and sub-Arctic seas. Annual Review of Marine Science 4:63-78.Grebmeier, J. M., J. E. Overland, S. E. Moore, E. V. Farley, E. C. Carmack, L. W. Cooper, K. E. Frey, J. H. Helle, F. A. McLaughlin, and S. L. McNutt. 2006. A major ecosystem shift in the northern Bering Sea. Science 311(5766):1461-1464.

Griffin, C. G., K. E. Frey, J. Rogan, and R. M. Holmes. 2011. Spatial and interannual variability of dissolved organic matter in the Kolyma River, East Siberia, observed using satellite imagery. Journal of Geophysical Research 116:G03018.

Grosse, G., V. Romanovsky, T. Jorgenson, K. W. Anthony, J. Brown, and P. P. Overduin. 2011. Vulnerability and feedbacks of permafrost to climate change. Eos, Transactions Americal Geophysical Union 92(9):73-74.

Gupta, P., D. Noone, J. Galewsky, C. Sweeney, and B. H. Vaughn. 2009. Demonstration of high-precision continuous measurements of water vapor isotopologues in laboratory and remote field deployments using wavelength-scanned cavity ring-down spectroscopy (WS-CRDS) technology. Rapid Communications in Mass Spectrometry 23(16):2534-2542.

Hacquebord, L., and D. Avango. 2009. Settlements in an Arctic resource frontier region. Arctic Anthropology 46(1-2):25-39.

Hamilton, L. 2009. Education, politics and opinions about climate change evidence for interaction effects. Climatic Change 104(2):231-242.

Hamilton, L., P. Bjerregaard, and B. Poppel. 2010. Health and population. In Arctic Social Indicators: A Follow-up to the Arctic Human Development Report, J. N. Larsen, P. Schweitzer, and G. Fondahl, eds. Copenhagen: Nordic Council of Ministers.

Hansen, M. C., R. S. DeFries, J. R. G. Townshend, and R. Sohlberg. 2000. Global land cover classification at 1 km spatial resolution using a classification tree approach. International Journal of Remote Sensing 21(6-7):1331-1364.

Hartman, C., and P. Johnson. 1978. Environmental Atlas of Alaska. Fairbanks: University of Alaska.

Harwood, D., F. Florindo, F. Talarico, R. Levy, G. Kuhn, T. Naish, F. Niessen, R. Powell, A. Pyne, and G. Wilson. 2009. Antarctic drilling recovers stratigraphic records from the continental margin. Eos, Transactions American Geophysical Union 90(11):90.

Heleniak, T. 2008. Changing settlement patterns across the Russian north at the turn of the millennium. In Russia's Northern Regions on the Edge: Communities, Industries and Populations from Murmansk to Magadan, V. Rautio and M. Tykkyläinen, eds. Helsinki: Kikimora.

Heleniak, T. 2009. Growth poles and ghost towns in the Russian far north. In Russia and the North, E. W. Rowe, ed. Ottawa, Canada: University of Ottawa Press.

Hofmann, E. E., P. H. Wiebe, D. P. Costa, and J. J. Torres. 2011. Introduction to understanding the linkages between Antarctic food webs and the environment: A synthesis of Southern Ocean GLOBEC studies. Deep Sea Research Part II: Topical Studies in Oceanography 58(13-16):1505-1507.

Holland, D. M., R. H. Thomas, B. De Young, M. H. Ribergaard, and B. Lyberth. 2008. Acceleration of Jakobshavn Isbræ triggered by warm subsurface ocean waters. Nature Geoscience 1(10):659-664.

Holliday, N. P., S. L. Hughes, A. Lavin, K. A. Mork, G. Nolan, W. Walcowski, and A. Breszczynska-Moller. 2007. The end of a trend? The progression of unusually warm and saline water from the eastern North Atlantic into the Arctic Ocean. CLIVAR Exchanges 12(1):19-20.

Holmes, R. M., J. W. McClelland, B. J. Peterson, S. E. Tank, E. Bulygina, T. I. Eglinton, V. V. Gordeev, T. Y. Gurtovaya, P. A. Raymond, D. J. Repeta, R. Staples, R. G. Striegl, A. V. Zhulidov, and S. A. Zimov. 2011. Seasonal and annual fluxes of nutrients and organic matter from large rivers to the Arctic Ocean and surrounding seas. Estuaries and Coasts 35:369-382.

Hovelsrud, G. K., and B. Smit. 2010. Community adaptation and vulnerability in Arctic regions. Dordrecht, The Netherlands: Springer.

Hovelsrud, G. K., I. Krupnik, and J. White. 2011. Human-based observing systems. In Understanding Earth's Polar Challenges: International Polar Year 2007-2008, I. Krupnik, I. Allison, R. Bell, P. Cutler, D. Hik, J. López-Martínez, V. Rachold, E. Sarukhanian, and C. Summerhayes, eds. Edmonton, Canada: CCI Press.

Huntington, H. P., S. Gearheard, and L. K. Holm. 2010. The power of multiple perspectives: Behind the scenes of the Siku-Inuit-Hila Project. In SIKU: Knowing Our Ice, I. Krupnik, C. Aporta, S. Gearheard, G. J. Laidler, and L. Kielsen Holm, eds. Dordrecht, The Netherlands: Springer.

Huntington, H. P., L. C. Hamilton, C. Nicolson, R. Brunner, A. Lynch, A. E. J. Ogilvie, and A. Voinov. 2007. Understanding the human dimensions of the rapidly changing Arctic system: Insights and approaches from five HARC projects. Regional Environmental Change 7(4):173-186.

IASC (International Arctic Science Committee). 2011. State of the Arctic Coast 2010: Scientific Review and Outlook. Geesthacht, Germany: International Arctic Science Committee.

ICSU (International Council for Science). 2004. Scientific Data and Information: A Report of the CSPR Assessment Panel. Paris: International Council for Science.

International Permafrost Association Standing Committee on Data Information and Communication. 2010. IPA-IPY 2007-2009 Thermal State of Permafrost (TSP) Snapshot Borehole Inventory. Boulder, CO: National Snow and Ice Data Center. Available at *http://nsidc.org/data/g02190.html*, accessed May 3, 2012.

IPCC (Intergovernmental Panel on Climate Change). 2007a. Climate Change 2007: Synthesis Report. Contribution of Working Groups I, II, and III to the Fourth Assessment Report of the Intergovernmental Panel on Climate Change, S. Solomon, D. Qin, M. Manning, Z. Chen, M. Marquis, K.B. Averyt, M. Tignor, and H. L. Miller, eds. Geneva, Switzerland: Intergovernmental Panel on Climate Change.

IPCC. 2007b. The Physical Science Basis. Contribution of Working Group I to the Fourth Assessment Report of the Intergovernmental Panel on Climate Change, S. Solomon, D. Qin, M. Manning, Z. Chen, M. Marquis, K. B. Averyt, M. Tignor, and H. L. Miller, eds. Cambridge, UK: Cambridge University Press.

IPCC. 2011. IPCC Special Report on Managing the Risks of Extreme Events and Disasters to Advance Climate Change Adaptation: Summary for Policymakers. Geneva, Switzerland: Intergovernmental Panel on Climate Change.

Isaksen, K., R. E. Benestad, C. Harris, and J. L. Sollid. 2007. Recent extreme near-surface permafrost temperatures on Svalbard in relation to future climate scenarios. Geophysical Research Letters 34:L17502.

Jacob, D. J., J. H. Crawford, H. Maring, A. D. Clarke, J. E. Dibb, L. K. Emmons, R. A. Ferrare, C. A. Hostetler, P. B. Russell, H. B. Singh, A. M. Thompson, G. E. Shaw, E. McCauley, J. R. Pederson, and J. A. Fisher. 2010. The Arctic Research of the Composition of the Troposphere from Aircraft and Satellites (ARCTAS) mission: Design, execution, and first results. Atmospheric Chemistry and Physics 10:5191-5212.

Jacobs, S. S., A. Jenkins, C. F. Guilivi, and P. Dutrieux. 2011. Stronger ocean circulation and increased melting under Pine Island Glacier ice shelf. Nature Geoscience 4:519-523.

Jenkins, A., P. Dutrieux, S. S. Jacobs, S. D. McPhail, J. R. Perrett, A. T.Webb, and D. White. 2010. Observations beneath Pine Island Glacier in West Antarctica and implications for its retreat. Nature Geocience 3(7):468-472.

Johnson, J. A., W. P. Mason, A. J. Shturmakov, A. Jay, W. P. Mason, and A. Shturmakov. 2007. A new 122 mm electromechanical drill for deep ice-sheet coring (DISC): 5. Experience during Greenland field testing. Annals of Glaciology 47(1):54-60.

Jorgenson, M. T., V. Romanovsky, J. Harden, Y. Shur, J. O'Donnell, E. A. G. Schuur, M. Kanevskiy, and S. Marchenko. 2010. Resilience and vulnerability of permafrost to climate change. Canadian Journal of Forest Research 40(7):1219-1236.

Joughin, I., S. B. Das, M. A. King, B. Smith, I. Howatt, and T. Moon. 2008. Seasonal speedup along the western flank of the Greenland ice sheet. Science 320(5877):781-783.

Joughin, I., B. E. Smith, and D. M. Holland. 2010. Sensitivity of 21st century sea level to ocean-induced thinning of Pine Island Glacier, Antarctica. Geophysical Research Letters 37:L20502.

Juday, G. P., V. Barber, T. S. Rupp, J. Zasada, and W. M. Wilmking 2003. A 200-year perspective of climate variability and the response of white spruce in interior Alaska. Pp 226-250 in Climate Variability and Ecosystem Response at Long-Term Ecological Research Sites, D. Greenland, D. G. Goodin, and R. C. Smith, eds. Oxford: Oxford University Press..

Kaiser, B., ed. 2010. Polar Science and Global Climate: An International Resource for Education and Outreach. London: Pearson.

Kang, S., L. Polvani, J. Fyfe, and M. Sigmond. 2011. Impact of polar ozone depletion on subtropical precipitation. Science 332(6032):951-954.

Karcher, M., F. Kauker, R. Gerdes, E. Hunke, and J. Zhang. 2007. On the dynamics of Atlantic Water circulation in the Arctic Ocean. Journal of Geophysical Research 112:C04S02.

Kasischke, E., S. Goetz, J. Kimball, and M. Mack. 2011. The Arctic-Boreal Vulnerability Experiment (ABoVE): A Concise Plan for a NASA-Sponsored Field Campaign. Washington, DC: National Aeronautics and Space Administration.

Kaufman, D. S., D. P. Schneider, N. P. McKay, C. M. Ammann, R. S. Bradley, K. R. Briffa, G. H. Miller, B. L. Otto-Bliesner, J. T. Overpeck, B. M. Vinther, and Arctic Lakes 2k Project Members. 2009. Recent warming reverses long-term Arctic cooling. Science 325(5945):1236-1239.

Kelly, B., A. Whiteley, and D. Tallmon. 2010. The Arctic melting pot. Nature 468(7326):891.

King, J. R., G. McFarlane, A. Hollowed, A. MacCall, and J. Polovina. 2005. Report of the Study Group on Fisheries and Ecosystem Responses to Recent Regime Shifts. Sidney, British Columbia: North Pacific Marine Science Organization.

Koenig, S. J., R. M. DeConto, and D. Pollard. 2011. Late Pliocene to Pleistocene sensitivity of the Greenland ice sheet in response to external forcing and internal feedbacks. Climate Dynamics 37(5-6):1247-1268.

Kofinas, G. P., F. S. Chapin, S. BurnSilver, J. I. Schmidt, N. L. Fresco, K. Kielland, S. Martin, A. Springsteen, and T. S. Rupp. 2010. Resilience of Athabascan subsistence systems to interior Alaska's changing climate. Canadian Journal of Forest Research 40(7):1347-1359.

Konstantinov, Y. 2010. Socioeconomic life of climate change: Extensivity in reindeer husbandry in relation to synergies between social and climate change (Kola Peninsula). Acta Borealia 27(1):44-65.

Krupnik, I., and G. K. Hovelsrud. 2011. Polar societies and social processes. In Understanding Earth's Polar Challenges: International Polar Year 2007-2008, I. Krupnik, I. Allison, R. Bell, P. Cutler, D. Hik, J. López-Martínez, V. Rachold, E. Sarukhanian, and C. Summerhayes, eds. Edmonton, Canada: CCI Press.

Krupnik, I., and W. Weyapuk. 2010. *Qanuq Ilitaavut*: "How we learned what we know" (Wales Inupiaq Sea Ice Dictionary). In SIKU: Knowing Our Ice, I. Krupnik, C. Aporta, S. Gearheard, G. J. Laidler, and L. Kielsen Holm, eds. Dordrecht, The Netherlands: Springer.

Krupnik, I., C. Aporta, S. Gearheard, G. L. Laidler, and L. Kielsen Holm, eds. 2010a. Power of multiple perspectives: Behind the scenes of the Siku-Inuit-Hila Project. In SIKU: Knowing Our Ice, I. Krupnik, C. Aporta, S. Gearheard, G. J. Laidler, and L. Kielsen Holm, eds. Dordrecht, The Netherlands: Springer.

Krupnik, I., C. Aporta, S. Gearheard, G. J. Laidler, and L. Kielsen Holm, eds. 2010b. SIKU: Knowing Our Ice. Documenting Inuit Sea Ice Knowledge and Use. Dordrecht, The Netherlands, and Heidelberg, Germany: Springer.

Krupnik, I., I. Allison, R. Bell, P. Cutler, D. Hik, J. López-Martínez, V. Rachold, E. Sarukhanian, and C. Summerhayes, eds. 2011. Understanding Earth's Polar Challenges: International Polar Year 2007-2008. Summary by the IPY Joint Committee. Edmonton, Canada: CCI Press.

Kruse, J. 2010. Sustainability from a local point of view: Alaska's North Slope and oil development. In The Political Economy of Northern Regional Development, G. Winther, ed. Copenhagen: TemaNord.

Kwok, R., and G. F. Cunningham. 2008. ICESat over Arctic sea ice: Estimation of snow depth and ice thickness. Journal of Geophysical Research 113:C08010.

Kwok, R., and D. A. Rothrock. 2009. Decline in Arctic sea ice thickness from submarine and ICESat records: 1958-2008. Geophysical Research Letters 36:L15501.

Langley, K., J. Kohler, K. Matsuoka, A. Sinisalo, T. Scambos, T. Neumann, A. Muto, J. G. Winther, and M. Albert. 2011. Recovery Lakes, East Antarctica: Radar assessment of sub-glacial water extent. Geophysical Research Letters 38:L05501.

Langway, C. 2008. History of the early polar ice cores. Cold Regions Science and Technology 52(2):101-117.

Larsen, J., P. Schweitzer, and G. Fondahl. 2010. Arctic Social Indicators. Arhus: TemaNord.

Laubjerg, M., and B. Petersson. 2010. Greenlandic adoptees' psychiartric inpatient contact. A comparative register-based study. Scandinavian Journal of Public Health 38:64-70.

Laundal, K. M., and N. Østgaard. 2009. Asymmetric auroral intensities in the Earth's Northern and Southern Hemispheres. Nature 460:491-493.

Launius, R. D., J. R. Fleming, and D. H. DeVorkin, eds. 2010. Globalizing Polar Science: Reconsidering the International Polar and Geophysical Years. New York: Palgrave Macmillan.

Lawrence, D. M., A. G. Slater, R. A. Tomas, M. M. Holland, and C. Deser. 2008. Accelerated Arctic land warming and permafrost degradation during rapid sea ice loss. Geophysical Research Letters 35:L11506.

Lee, S. H., H. M. Joo, M. S. Yun, and T. E. Whitledge. 2011. Recent phytoplankton productivity of the northern Bering Sea during early summer in 2007. Polar Biology 35(1):83-89.

Leiserowitz, E. W., E. W. Maibach, C. Roser-Renouf, N. Smith, and E. Dawson. In press. Climategate, public opinion, and the loss of trust. American Behavioral Scientist.

Levintova, M., W. I. Zapol, and E. Engmann, eds. 2010. Behavioral and Mental Health Research in the Arctic: Strategy Setting Meeting. Circumpolar Health Supplements No. 5. Oulu, Finland: International Association of Circumpolar Health Publishers.Liao, J., L. G. Huey, D. J. Tanner, N. Brough, S. Brooks, J. E. Dibb, J. Stutz, J. L. Thomas, B. Lefer, C. Haman, and K. Gorham. 2011. Observations of hydroxyl and peroxy radical and the impact of BrO at Summit, Greenland in 2007 and 2008. Atmospheric Chemistry and Physics 11:8577-8591.

Liggett, D. and G. Steel. 2011. Adding "value" to Antarctic research, management and policy: A project-driven approach towards creating a community of Antarctic social and human scientists. Presented at the 7th International Congress of Arctic Social Sciences (ICASS VII), Akureyri, Iceland, 22-26 June 2011.

Lovecraft, A. L., and H. Eicken. 2011. North by 2020: Perspectives on Alaska's Changing Social-Ecological Systems. Fairbanks: University of Alaska Press.

Lozhkin, A., and P. Anderson. 2011. Forest or no forest: Implications of the vegetation record for climatic stability in Western Beringia during Oxygen Isotope Stage 3. Quaternary Science Reviews 30(17-18):2160-2181.

Mack, M. C., M. S. Bret-Harte, T. N. Hollingsworth, R. R. Jandt, E. A. G. Schuur, G. R. Shaver, and D. L. Verbyla. 2011. Carbon loss from an unprecedented Arctic tundra wildfire. Nature 475(7357):489-492.

Manabe, S., and R. T. Wetherald. 1975. The effects of changing the solar constant on the climate of a general circulation model. Journal of Atmospheric Sciences 32(11):2044-2059.

Mann, M. E., Z. Zhang, M. K. Hughes, R. S. Bradley, S. K. Miller, S. Rutherford, and F. Ni. 2008. Proxy-based reconstructions of hemispheric and global surface temperature variations over the past two millennia. Proceedings of the National Academy of Sciences of the United States of America 105(36):13252-13257.

Marchant, D. R., C. C. Swisher, D. R. Lux, D. P. West, and G. H. Denton. 1993. Pliocene paleoclimate and East Antarctic ice-sheet history from surficial ash deposits. Science 260(5108):667-670.

Maslanik, J. A., C. Fowler, J. Stroeve, S. Drobot, J. Zwally, D. Yi, and W. Emery. 2007. A younger, thinner Arctic ice cover: Increased potential for rapid, extensive sea-ice loss. Geophysical Research Letters 34:L24501.

Maslowski, W., D. C. Marble, W. Walczowski, and A. J. Semtner. 2001. On large scale shifts in the Arctic Ocean and sea ice conditions during 1979-1998. Annals of Glaciology 33:545-550.

Mason, W. A. S., J. Johnson, and S. Haman. 2007. A new 122mm electromechanical drill for deep ice-sheet coring (DISC): 2. Mechanical design. Annals of Glaciology 47:35-40.

Mathis, J. T., D. A. Hansell, and N. R. Bates. 2009. Interannual variability of dissolved inorganic carbon distribution and net community production during the Western Arctic Shelf-Basin Interactions Project. Deep Sea Research II 56(7):1213-1222.

Matsuno, K., and A. Yamaguchi. 2010. Abundance and biomass of mesozooplankton along north-south transects (165°E and 165°W) in summer in the North Pacific: An analysis with an optical plankton counter. Plankton and Benthos Research 5(4):123-130.

Mazloff, M. P. H., and C. Wunsch. 2010. An eddy-permitting Southern Ocean state estimate. Journal of Geophysical Research 40:880-899.

McGregor, R. V. 2010. Guidelines for development and management of transportation infrastructure in permafrost regions: Ottawa: Transportation Association of Canada.

McNeil, B., and R. Matear. 2008. Southern Ocean acidification: A tipping point at 450-ppm atmospheric $CO_2$. Proceedings of the National Academy of Sciences of the United States of America 105(48):18860-18864.

Melles, M., J. Brigham-Grette, P. Minyuk, C. Koeberl, A. Andrej, T. Cook, G. Fedorov, C. Gebhardt, E. Haltia-Hovi, M. Kukkonen, N. Nowaczyk, G. Schwamborn, V. Wennrich, and El'gygytgyn Scientific Party. 2011. The Lake El'gygytgyn Scientific Drilling Project: Conquering Arctic challenges in continental drilling. Scientific Drilling 11:29-40.

Mende, S. B., W. Rachelson, R. Sterling, H. U. Frey, S. E. Harris, S. McBride, T. J. Rosenberg, D. Detrick, J. L. Doolittle, M. Engebretson, U. Inan, J. W. Labelle, L. J. Lanzerotti, and A. T. Weatherwax. 2009. Observations of Earth space by self-powered stations in Antarctica. Review of Scientific Instruments 80:124501.

Meredith, M., K. Nicholls, I. Renfrew, L. Boehme, M. Biuw, and M. Fedak. 2011. Seasonal evolution of the upper-ocean adjacent to the South Orkney Islands, Southern Ocean: Results from a "lazy biological mooring." Deep Sea Research Part II: Topical Studies in Oceanography 58:1569-1579.

Miller, G., R. Alley, J. Brigham-Grette, J. Fitzpatrick, L. Polyak, M. Serreze, and J. W. C. White. 2010a. Arctic amplification: Can the past constrain the future? Quaternary Science Reviews 29(15-16):1779-1790.

Miller, G. H., J. Brigham-Grette, R. B. Alley, L. Anderson, H. A. Bauch, M. S. V. Douglas, M. E. Edwards, S. A. Elias, B. P. Finney, J. J. Fitzpatrick, S. V. Funder, T. D. Herbert, L. D. Hinzman, D. S. Kaufman, G. M. MacDonald, L. Polyak, A. Robock, M. C. Serreze, J. P. Smol, R. Spielhagen, J. W. C. White, A. P. Wolfe, and E. W. Wolff. 2010b. Temperature and precipitation history of the Arctic. Quaternary Science Reviews 29:1679-1715.

Moberg, A., D. M. Sonechkin, K. Holmgren, N. M. Datsenko, and W. Karlén. 2008. Highly variable Northern Hemisphere temperatures reconstructed from low- and high-resolution proxy data. Nature 433:613-617.

Mohatt, G. V., J. Klejka, B. B. Boyer, B. R. Luick, C. Lardon, R. Plaetke, C. Wolsko, S. E. Hopkins, M. Dondanville, J. Herron, E. Ruppert, A. Bersamin, E. M. Orr, and The CANHR Research Team. 2007. The Center for Alaska Native Health Research Study: A community-based participatory research study of obesity and chronic disease-related protective and risk factors. International Journal of Circumpolar Health 66(1):8-18.

Moline, M. A., N. J. Karnovsky, Z. Brown, G. J. Divoky, T. R. Frazer, C. A. Jacoby, J. J. Torres, and W. R. Fraser. 2008. High latitude changes in ice dynamics and their impact on polar marine ecosystems. Annals of the New York Academy of Sciences 1134:267-319.

Morison, J., K. Aagaard, and M. Steele. 2000. Recent environmental changes in the Arctic: A review. Arctic 53(4):359-371.

Mortensen, N. B., P. J. Sendelbach, and A. J. Shturmakov. 2007. A new 122 mm electromechanical drill for deep ice-sheet coring (DISC): 3. Control, electrical and electronics design. Annals of Glaciology 47:41-50.

Muckle, G., E. Dewailly, P. Ayotte, P. Plusquellec, J. L. Jacobson, S. W. Jacobson, S.W. C. Bastien, D. Saint-Amour, and M. Boivin. 2009. Nunavik cohort study on exposure to environmental contaminants and child development. Pp. 3-7 in Synopsis of Research Conducted Under the 2008-2009 Northern Contaminants Program, S. Smith, J. Stow, and J. Edwards, eds. Ottawa, Canada: Department of Indian Affairs and Northern Development.

Musko, S. B., C. R. Clauer, A. J. Ridley, and K. L. Arnett. 2009. Autonomous low-power magnetic data collection platform to enable remote high latitude array deployment. Review of Scientific Instruments 80(4):044501.

Muto, A., T. A. Scambos, K. Steffen, A. G. Slater, and G. D. Clow. 2011. Recent surface temperature trends in the interior of East Antarctica from borehole firn temperature measurements and geophysical inverse methods. Geophysical Research Letters 38:L15502.

Naish, T., R. Powell, R.Levy, L. Krissek, F. Niessen, M. Pompilio, R. Scherer, F. Talarico, G. Wilson, T. Wilson, P. Barrett, G. Browne, L. Carter, R. Cody, E. Cowan, J. Crampton, R. DeConto, G. Dunbar, N. Dunbar, F. Florindo, C. Gebhardt, I. Graham, M. Hannah, D. Harwood, D. Hansaraj, S. Henrys, D. Helling, G. Kuhn, P. Kyle, A. Läufer, P. Maffioli, D. Magens, K. Mandernack, W. McIntosh, R. McKay, C. Millan, R. Morin, C. Ohneiser, T. Paulsen, D. Persico, D. Pollard, J. Reed, J. Ross, I. Raine, D. Schmitt, L. Sagnotti, C. Sjunneskog, P. Strong, M. Taviani, S. Vogel, T. Wilch, T. Williams, and D. Winter. 2009. Obliquity-paced Pliocene West Antarctic ice sheet oscillations. Nature 458:322-328.

NASA (National Aeronautics and Space Administration) and University of Maryland. 2002. MODIS Hotspot/Active Fire Detections. Data set. MODIS Rapid Response Project, NASA/GSFC, University of Maryland, Fire Information for Resource Management System. Available at *http://earthdata.nasa.gov/data/nrt-data/firms*, accessed June 27, 2012.

National Commission on the BP Deepwater Horizon Oil Spill and Offshore Drilling. 2011. Deep Water: The Gulf Oil Disaster and the Future of Offshore Drilling. Report to the President. Washington, DC: U.S. Government Printing Office.

Nicholls, R. J., and A. Cazenave. 2010. Sea-level rise and its impact on coastal zones. Science 328:1517-1520.

Nicholls, R. J., S. Hanson, C. Herweijer, N. Patmore, S. Hallegatte, J. Corfee-Morlot, J. Château, and R. Muir-Wood. 2007. Ranking of the World's Cities Most Exposed to Coastal Flooding Today and in the Future: Executive Summary. Paris: Organisation for Economic Co-operation and Development.

Nolan, M., A. Arendt, B. Rabus, and L. Hinzman. 2005. Volume change of McCall Glacier, Arctic Alaska, from 1956 to 2003. Annals of Glaciology 42:409-416.

NRC (National Research Council). 2004. A Vision for the International Polar Year 2007-2008. Washington, DC: National Academies Press.

NRC. 2005. International Polar Year 2007-2008: Report of the Implementation Workshop. Washington, DC: National Academies Press.

NRC. 2008. Martha Muse Prize for Science and Policy in Antarctica: An International Polar Year Legacy. Washington, DC: National Academies Press.

NRC. 2010a. Advancing the Science of Climate Change. Washington, DC: National Academies Press.

NRC. 2010b. Ocean Acidification: A National Strategy to Meet the Challenges of a Changing Ocean. Washington, DC: National Academies Press.

NRC. 2011a. Climate Stablization Targets: Emissions, Concentrations, and Impacts over Decades to Millennia. Washington, DC: National Academies Press.

NRC. 2011b. Future Science Opportunities in Antarctica and the Southern Ocean. Washington, DC: National Academies Press.

NSF (National Science Foundation). 2007. NSF Merit Review Broader Impacts Criterion: Representative Activities. Available at *http://www.nsf.gov/pubs/gpg/broaderimpacts.pdf*, accessed April 30, 2012.

O'Donnell, R., N. Lewis, S. McIntyre, and J. Condon. 2010. Improved methods for PCA-based reconstructions: Case study using the Steig et al. (2009) Antarctic temperature reconstruction. Journal of Climate 24:2099-2115.

Oskal, A., J. M. Turi, S. D. Mathiesen, and P. Burgess, eds. 2009. EALÁT. Reindeer Herders Voice: Reindeer Herding, Traditional Knowledge and Adaptation to Climate Change and Loss of Grazing Lands. Report No. 2. Kautokeino: International Centre for Reindeer Husbandry.

Otto-Bliesner, B. L., S. J. Marshall, J. T. Overpeck, G. H. Miller, A. Hu, and CAPE Last Interglacial Project members. 2006. Simulating Arctic climate warmth and icefield retreat in the last interglaciation. Science 311(5768):1751-1753.

Overland, J. E. 2011. Potential Arctic change through climate amplification processes. Oceanography 24(3):176-185.

Overland, J. E., and M. Wang. 2009. Large-scale atmospheric circulation changes are associated with the recent loss of Arctic sea ice. International Journal of Climatology 29(15):2348-2352.

Overland, J. E., and M. Wang. 2010. Large-scale atmospheric circulation changes are associated with the recent loss of Arctic sea ice. Tellus, Series A: Dynamic Meteorology and Oceanography 62(1):1-9.

Overland, J. E., L. Bengtsson, R. Przybylak, and J. Walsh. 2007. Atmosphere. In Arctic Report Card 2007, J. Richter-Menge, J. Overland, E. Hanna, M. J. J. E. Loonen, A. Proshutinsky, V. Romanovsky, D. Russell, R. Van Bogaert, R. Armstrong, L. Bengtsson, J. Box, T. V. Callaghan, M. De Dapper, B. Ebbinge, O. Grau, M. Hallinger, L. D. Hinzman, P. Huybrechts, G. J. Jia, C. Jonasson, J. Morison, S. Nghiem, N. Oberman, D. Perovich, R. Przybylak, I. Rigor, A. Shiklomanov, D. Walker, J. Walsh, and C. Zöckler, eds. Available at *http://www.arctic.noaa.gov/report07/atmosphere.html*, accessed April 30, 2012.

Overland, J. E., J. Turner, J. Francis, N. Gillett, G. Marshall, and M. Tjernström. 2008. The Arctic and Antarctic: Two faces of climate change. Eos, Transactions American Geophysical Union 89(19):177-178.

Overpeck, J. 2005. Arctic system on trajectory to new, seasonally ice-free state. Eos, Transactions American Geophysical Union 86(34):309-313.

Overpeck, J., K. Hughen, D. Hardy, R. Bradley, R. Case, M. Douglas, B. Finney, K. Gajewski, G. Jacoby, A. Jennings, S. Lamoureux, A. Lasca, G. MacDonald, J. Moore, M. Retelle, S. Smith, A. Wolfe, and G. Zielinski. 1997. Arctic environmental change of the last four centuries. Science 278(5341):1251-1256.

Parkinson, A. 2010. Arctic Health Initiative. Circumpolar Health Supplements No. 6. Oulu, Finland: International Association of Circumpolar Health Publishers.

Parkinson, A., M. Bruce, T. Zulz, and The International Circumpolar Surveillance Steering Committee. 2008. International Circumpolar Surveillance, an Arctic network for the surveillance of infectious diseases. Emerging Infectious Diseases 14(1), available at *http://wwwnc.cdc.gov/eid/article/14/1/07-0717.htm*.

Parsekian, A. D., B. M. Jones, M. Jones, G. Grosse, K. Walter Anthony, and L. Slater. 2011. Expansion rate and geometry of floating vegetation mats on the margins of thermokarst lakes, northern Seward Peninsula, Alaska, USA. Earth Surface Processes and Landforms 36(14):1889-1897.

Parsons, M. A., T. deBruin, S. Tomlinson, H. Campbell, Ø. Godøy, and J. LeClert. 2011. The state of polar data: The IPY experience. In Understanding Earth's Polar Challenges: International Polar Year 2007-2008, I. Krupnik, I. Allison, R. Bell, P. Cutler, D. Hik, J. López-Martínez, V. Rachold, E. Sarukhanian, and C. Summerhayes, eds. Edmonton, Canada: CCI Press.

Payne, J. F. 2010. 2010 Report to Congress, North Slope Science Initiative Organization and Accomplishments. Anchorage, AK: North Slope Science Initiative Oversight Group.

Perovich, D. K., and J. A. Richter-Menge. 2009. Loss of sea ice in the Arctic. Annual Review of Marine Science 1:417-441.

Perovich, D. K., B. Light, H. Eicken, K. F. Jones, K. Runciman, and S. V. Nghiem. 2007. Increasing solar heating of the Arctic Ocean and adjacent seas, 1979-2005: Attribution and role in the ice-albedo feedback. Geophysical Research Letters 34:L19505.

Perry, D., and E. Gyllenhaal. 2010. Summative Evaluation of POLAR-PALOOZA! for Geoff Haines-Stiles Productions, Inc. Chicago: Selinda Research Associates, Inc. Available at *http://staging.informalscience.org/reports/0000/0231/PPZA_Final_Report_92.pdf*, accessed April 30, 2012.

Pfirman, S., R. Bell, M. Turrin, and P. Maru. 2004. Bridging the Poles: Education Linked with Research. A Report on the Workshop. New York: Lamont-Doherty Earth Observatory.

Pfirman, S., B. Tremblay, C. Fowler, and R. Newton. 2009. The last Arctic sea ice refuge. The Circle 4:6-8.

Piñones, A., E. Hofmann, M. Dinniman, and J. Klinck. 2011. Lagrangian simulation of transport pathways and residence times along the western Antarctic Peninsula. Deep Sea Research Part II: Topical Studies in Oceanography 58(13-16):1524-1539.

Pollard, D., and R. M. DeConto. 2009. Modelling West Antarctic ice sheet growth and collapse through the past five million years. Nature 458(7236):329-332.

Polyak, L., R. Alley, J. Andrews, J. Brigham-Grette, T. Cronin, D. Darby, A. Dyke, J. Fitzpatrick, S. Funder, M. Holland, A. Jennings, G. Miller, M. O'Regan, J. Savelle, M. Serreze, K. S. John, J. White, and E. Wolff. 2010. History of sea ice in the Arctic. Quaternary Science Reviews 29(15-16):1757-1778.

Polyakov, I. V., L. Timokhov, I. Dmitrenko, V. Ivanov, H. Simmons, A. Beszczynska-Möller, R. Dickson, E. Fahrbach, L. Fortier, J.-C. Gascard, J. Hölemann, N. P. Holliday, E. Hansen, C. Mauritzen, J. Piechura, R. Pickart, U. Schauer, W. Walczowski, and M. Steele. 2007. Observational program tracks Arctic Ocean transition to a warmer state. Eos, Transactions American Geophysical Union 88(40):398-399.

Polyakov, I. V., A. Pnyushkov, R. Rember, V. Ivanov, Y.-D. Lenn, L. Padman, and E. Carmack. 2011. Mooring-based observations of double-diffusive staircases over the Laptev Sea slope. Journal of Physical Oceanography 42(1):95-109.

Pritchard, H. D., R. J. Arthern, D. G. Vaughan, and L. A. Edwards. 2009. Extensive dynamic thinning on the margins of the Greenland and Antarctic ice sheets. Nature 461(7266):971-975.

Rand, K. M., and E. A. Logerwell. 2011. The first demersal trawl survey of benthic fish and invertebrates in the Beaufort Sea since the late 1970s. Polar Biology 34(4):475-488.

Rapley, C., and R. Bell. 2004. A Framework for the International Polar Year 2007-2008. Paris: International Council for Science.

Rawlins, M. A., M. Steele, M. M. Holland, J. C. Adam, J. E. Cherry, J. A. Francis, P. Y. Groisman, L. D. Hinzman, T. G. Huntington, D. L. Kane, J. S. Kimball, R. Kwok, R. B. Lammers, C. M. Lee, D. P. Lettenmaier, K. C. McDonald, E. Podest, J. W. Pundsack, B. Rudels, M. C. Serreze, A. Shiklomano, O. Skagseth, T. J. Troy, C. J. Vorosmarty, M. Wensnahan, E. F. Wood, R. Woodgate, D. Q. Yang, K. Zhang, and T. J. Zhang. 2010. Analysis of the Arctic system for freshwater cycle intensification: Observations and expectations. Journal of Climate 23(21):5715-5737.

Raymo, M., J. Mitrovica, M. J. O'Leary, R. M. DeConto, and P. J. Hearty. 2011. Departures from eustasy in Pliocene sea level records. Nature Geoscience 4:328-332.

Revkin, A. C. 2007. 2-year study of polar changes set to begin. The New York Times, February 26, 2007.

Rignot, E., G. Casassa, P. Gogineni, W. Krabill, A. Rivera, and R. Thomas. 2004. Accelerated ice discharge from the Antarctic Peninsula following the collapse of Larsen B Ice Shelf. Geophysical Research Letters 31(18):L18401.

Rignot, E., M. Koppes, and I. Velicogna. 2010. Rapid submarine melting of the calving faces of West Greenland glaciers. Nature Geoscience 3:187-191.

Rignot, E., J. Mouginot, and B. Scheuchl. 2011. Ice flow of the Antarctic ice sheet. Science 333(6048):1427-1430.

Rigor, I., and J. M. Wallace. 2004. Variations in the age of Arctic sea-ice and summer sea-ice extent. Geophysical Research Letters 31:L09401.

Rink, E., D. Gensink-Law, R. Montgomery-Andersen, G. Mulvad, and A. Koch. 2009. The practical application of community based participatory reseraarch in Greenland: Initial experiences of the Greenland Sexual Health Study. International Journal of Circumpolar Health 68(4):405-413.

Rintoul, S. R., M. D. Sparrow, M. P. Meredith, V. Wadley, K. Speer, E. Hofmann, C. Summerhayes, E. Urban, and E. R. Bellerby. 2011. The Southern Ocean Observing System: Initial science and implementation strategy. Cambridge, UK: Scientific Committee on Antarctic Research.

Roach, J., B. Griffith, D. Verbyla, and J. Jones. 2011. Mechanisms influencing changes in lake area in Alaskan boreal forest. Global Change Biology 17(8):2567-2583.

Roberts, A., J. Cassano, R. Döscher, L. Hinzman, M. Holland, H. Mitsudera, A. Sumi, and J. E. Walsh. 2010. A Science Plan for Regional Arctic System Modeling: A Report by the Arctic Research Community for the National Science Foundation Office of Polar Programs. Fairbanks: International Arctic Research Center, University of Alaska.

Rode, K. D., S. C. Amstrup, and E. V. Regehr. 2010. Reduced body size and cub recruitment in polar bears associated with sea ice decline. Ecological Applications 20:768-782.

Romanovsky, V. E., N. G. Oberman, D. S. Drozdov, G. V. Malkova, N. G. Moskalenko, A. A. Vasiliev, D. O. Sergeev, and D. G.Vaughan. 2008. Thermal state and fate of permafrost in Russia: First results of IPY (plenary paper). Presented at Ninth International Conference on Permafrost, Fairbanks, Alaska.

Romanovsky, V. E., D. S. Drozdov, N. G. Oberman, G. V. Malkova, A. L. Kholodov, S. S. Marchenko, N. G. Moskalenko, D. O. Sergeev, N. G. Ukraintseva, A. A. Abramov, D. A. Gilichinsky, and A. A. Vasiliev. 2010. Thermal state of permafrost in Russia. Permafrost and Periglacial Processes 21(2):136-155.

Rosenzweig, C., W. Solecki, A. DeGaetano, S. Hassol, P. Grabhorn, M. O'Grady, D. Bader, F. Buonaiuto, J. Comstock, S. A. Hammer, R. Horton, K. Jacob, P. L. Kinney, Y. Klein, R. Leichenko, D. C. Major, A. McDonald, R. S. Ostfeld, M. Panero, L. Parshall, L. Patrick, R. Schneider, P. Sheffield, L. Tryhorn, P. Vancura, and D. W. Wolfe. 2011. Responding to Climate Change in New York State: Synthesis Report. New York: New York State Energy Research and Development Authority.

Sakamoto, T., Y. Tanaka, J. Simonetti, C. Osiowy, M. L. Børresen, A. Koch, F. Kubanov, M. Sugiyama, G. Y. Minuk, B. J. McMahon, and M. Mizokami. 2007. Classification of hepatitis B virus genotype B into two major forms based on characterization of a novel subgenotype in the Arctic indigenous populations. Journal of Infecitons Diseases 196:1487-1492.

Salawitch, R. J., T. Canty, T. Kurosu, K. Chance, Q. Liang, A. da Silva, S. Pawson, J. E. Nielsen, J. M. Rodriguez, P. K. Bhartia, X. Liu, L. G. Huey, J. Liao, R. E. Stickel, D. J. Tanner, J. E. Dibb, W. R. Simpson, D. Donohoue, A. Weinheimer, F. Flocke, D. Knapp, D. Montzka, J. A. Neuman, J. B. Nowak, T. B. Ryerson, S. Oltmans, D. R. Blake, E. L. Atlas, D. E. Kinnison, S. Tilmes, L. L. Pan, F. Hendrick, M. V. Roozendael, K. Kreher, P. V. Johnston, R. S. Gao, B. Johnson, T. P. Bui, G. Chen, R. B. Pierce, J. H. Crawford, and D. J. Jacob. 2011. A new interpretation of total column BrO during Arctic spring. Geophysical Research Letters 37:L21805.

Salmon, R. A., D. J. Carlson, S. Zicus, M. Pauls, J. Baeseman, E. B. Sparrow, K. Edwards, M. H. Almeida, L. T. Huffman, T. Kolset, R. J. H. Malherbe, M. S. McCaffrey, N. A. L. Munro, J. de Pomereu, J. Provencher, K. A. Rahman-Sinclair, and M. Raymond. 2011. Education, outreach and communication during the International Polar Year 2007-2008: Stimulating a global polar community. Polar Journal 1(2):265-285.

Scafetta, B., and B. J. West. 2005. Estimate solar contribution to the global surface warming using the ACRIM TSI satellite composite. Geophysical Research Letters 32:L18713.

Scambos, T. A., J. A. Bohlander, C. A. Shuman, and P. Skvarca. 2004. Glacier acceleration and thinning after ice shelf collapse in the Larsen B embayment, Antarctica. Geophysical Research Letters 31:L18402.

Scambos, T. A., M. Frezzotti, T. Haran, J. Bohlander, J. T. M. Lenaerts, M. R. van den Broeke, K. Jezek, D. Long, S. Urbini, K. Farness, T. Neumann, M. Albert, and J. Gunnar-Winther. 2011. Extent of zero-accumulation "wind glaze" areas on the East Antarctic plateau: Implications for continental ice mass balance. Journal of Glaciology, in review.

Scherer, R. P., S. M. Bohaty, R. B. Dunbar, O. Esper, J.-A. Flores, R. Gersonde, D. M. Harwood, A. P. Roberts, and M. Taviani. 2008. Antarctic records of precession-paced insolation-driven warming during early Pleistocene Marine Isotope Stage 31. Geophysical Research Letters 35:L03505.

Schneider, D. P., C. Deser, and Y. Okumura. 2011. An assessment and interpretation of the observed warming of West Antarctica in the austral spring. Climatic Dynamics 38(1/2):323.

Schuur, E. A. G., J. G. Vogel, K. G. Crummer, H. Lee, J. O. Sickman, and T. E. Osterkamp. 2009. The effect of permafrost thaw on old carbon release and net carbon exchange from tundra. Nature 459(7246):556-559.

Schweitzer, P. P., E. Khlinovskaya-Rockhill, F. Stammler and T. Heleniak, eds. In press. Moved by the State: Population Movements and Agency in the Circumpolar North and Other Remote Regions. New York: Berghahn Press.

Science and Applications Working Group. 1979. Ice and Climate Experiment: Report of Science and Applications Working Group, Goddard Space Flight Center. Greenbelt, MD: NASA Goddard Space Flight Center.

Secretariat of the Antarctic Treaty. 2009. Final Report of the Thirty-Second Antarctic Treaty Consultative Meeting, Baltimore, Maryland, April 6-17, 2009. Buenos Aires, Argentina: Secretariat of the Antarctic Treaty.

Seppälä, A., C. E. Randall, M. A. Clilverd, E. Rozanov, and C. J. Rodger. 2009. Geomagnetic activity and polar surface air temperature variability. Journal of Geophysical Research 114:A10312.

Serreze, M. C., and R. G. Barry. 2011. Processes and impacts of Arctic amplification: A research synthesis. Global and Planetary Change 77:85-96.

Serreze, M. C., M. M. Holland, and J. Stroeve. 2007. Perspectives on the Arctic's shrinking sea-ice cover. Science 315(5818):1533-1536.

Serreze, M. C., A. P. Barrett, J. C. Stroeve, D. N. Kindig, and M. M. Holland. 2009. The emergence of surface-based Arctic amplification. The Cryosphere 3:11-19.

Shadian, J. M., and M. Tennberg, eds. 2009. Legacies and Change in Polar Sciences: Historical, Legal, and Political Reflections on International Polar Year. Farnham, UK: Ashgate.

Shindell, D., M. Schulz, Y. Ming, T. Takemura, G. Faluvegi, and V. Ramaswamy. 2010. Spatial scales of climate response to inhomogeneous radiative forcing. Journal of Geophysical Research 115:D19110.

Shturmakov, A. J., D. A. Lebar, W. P. Mason, and C. R. Bentley. 2007. A new 122 mm electromechanical drill for deep ice-sheet coring (DISC): 1. Design concepts. Annals of Glaciology 47(1):28-34.

Sigmond, M., M. C. Reader, J. C. Fyfe, and N. P. Gillett. 2011. Drivers of past and future Southern Ocean change: Stratospheric ozone versus greenhouse gas impacts. Geophysical Research Letters 38:L12601.

Smith, S. L., V. E. Romanovsky, A. G. Lewkowicz, C. R. Burn, M. Allard, G. D. Clow, K. Yoshikawa, and J. Throop. 2010. Thermal state of permafrost in North America: A contribution to the International Polar Year. Permafrost and Periglacial Processes 21(2):117-135.

Stammerjohn, S. E., D. G. Martinson, R. C. Smith, X. Yuan, and D. Rind. 2008. Trends in Antarctic annual sea ice retreat and advance and their relation to El Niño-Southern Oscillation and Southern Annular Mode variability. Journal of Geophysical Research 113:C03S90.

Stearns, L., B. Smith, and G. Hamilton. 2008. Increased flow speed on a large East Antarctic outlet glacier caused by subglacial floods. Nature Geoscience 1(12):827-831.

Steele, M., W. Ermold, and J. L. Zhang. 2008. Arctic Ocean surface warming trends over the past 100 years. Geophysical Research Letters 35:L02614.

Steffensen, J. P., K. K. Andersen, M. Bigler, H. B. Clausen, D. Dahl-Jensen, H. Fischer, K. Goto-Azuma, M. Hansson, S. J. Johnsen, J. Jouzel, V. Masson-Delmotte, T. Popp, S. O. Rasmussen, R. Röthlisberger, U. Ruth, B. Stauffer, M. L. S. Andersen, Á. E. Sveinbjörnsdóttir, A. Svensson, and J. W. C. White. 2008. High-resolution Greenland ice core data show abrupt climate change happens in few years. Science 321(5889):680-684.

Steig, E. J., D. P. Schneider, S. D. Rutherford, M. E. Mann, J. C. Comiso, and D. T. Shindell 2009. Warming of the Antarctic ice-sheet surface since the 1957 International Geophysical Year. Nature Geoscience 457:459-462.

Steig, E. J., Q. Ding, D. S. Battisti, and A. Jenkins. 2011. Tropical forcing of Circumpolar Deep Water inflow and outlet glacier thinning in the Amundsen Sea embayment, West Antarctica. Annals of Glaciology 53(60).19-28

Stephenson, S. R., L. C. Smith, and J. A. Agnew. 2011. Divergent long-term trajectories of human access to the Arctic. Nature Climate Change 1(3):156-160.

Stokes, D. 1997. Pasteur's Quadrant: Basic science and technological innovation. Washington, DC: Brookings Institution Press.

Straneo, F., G. Hamilton, D. Sutherland, L. Stearns, F. Davidson, M. Hammill, G. Stenson, and A. Rosing-Asvid. 2010. Rapid circulation of warm subtropical waters in a major glacial fjord in East Greenland. Nature Geoscience 3:182-186.

Stroeve, J., M. M. Holland, W. Meier, T. Scambos, and M. Serreze. 2007. Arctic sea ice decline: Faster than forecast. Geophysical Research Letters 34(9):L09501.

Stroeve, J., M. Serreze, S. Drobot, S. Gearhear, M. Holland, J. Maslanik, W. Meier, and T. Scambos. 2008. Arctic sea ice extent plummets in 2007. Eos, Transactions American Geophysical Union 89(2):13-14.

Stroeve, J., M. Serreze, M. Holland, J. Kay, J. Maslanik, and A. Barrett. 2011. The Arctic's rapidly shrinking sea ice cover: A research synthesis. Climatic Change, doi: 10.1007/s10584-011-0101-1.

Stutz, J., J. L. Thomas, S. C. Hurlock, M. Schneider, R. von Glasow, M. Piot, K. Gorham, J. F. Burkhart, L. Ziemba, J. E. Dibb, and B. L. Lefer. 2011. Longpath DOAS observations of surface BrO at Summit, Greenland. Atmospheric Chemistry and Physics 11:9899-9910.

Takahashi, T., S. C. Sutherland, R. Wanninkhof, C. Sweeney, R. Feely, D. Chipman, B. Hales, G. Friederich, F. Chavez, C. Sabine, A. Watson, D. Bakker, U. Schuster, N. Metzl, H. Yoshikawa-Inoue, M. Ishii, T. Midorikawa, Y. Nojiri, A. Körtzinger, T. Steinhoff, M. Hoppema, J. Olafsson, T. Arnarson, B.Tilbrook, T. Johannessen, A. Olsen, R. Bellerby, C. S.Wong, B. Delille, N. R. Bates, and H. de Baar. 2009. Climatological mean and decadal change in surface ocean $pCO_2$, and net sea-air $CO_2$ flux over the global oceans. Deep Sea Research Part II: Topical Studies in Oceanography 56(8-10):554-577.

Tarnocai, C., J. G. Canadell, E. A. G. Schuur, P. Kuhry, G. Mazhitova, and S. Zimov. 2009. Soil organic carbon pools in the northern circumpolar permafrost region. Global Biochemical Cycles 23:GB2023.

Thomas, J. L., J. Stutz, B. Lefer, L. G. Huey, K. Toyota, J. E. Dibb, and R. von Glasow. 2011. Modeling chemistry in and above snow at Summit, Greenland, Part 1: Model description and results. Atmospheric Chemistry and Physics 11:4899-4914.

Thompson, D. W. J., and S. Solomon. 2002. Interpretation of recent Southern Hemisphere climate change. Science 296(5569):895-899.

Thompson, D. W. J., S. Solomon, P. J. Kushner, M. H. England, K. M. Grise, and D. J. Karoly. 2011. Signatures of the Antarctic ozone hole in Southern Hemisphere surface climate change. Nature Geoscience 4:741-749.

Thurairajah, B., R. L. Collins, V. L. Harvey, R. S. Lieberman, M. Gerding, K. Mizutani, and J. M. Livingston. 2010. Gravity wave activity in the Arctic stratosphere and mesosphere during the 2007-2008 and 2008-2009 stratospheric sudden warming events. Journal of Geophysical Research 115(17):D00N06.

Treadwell, M. 2009. IPY heightens attention to Arctic issues. Witness the Arctic 13(2). Available at *http://www.arcus.org/witness-the-arctic/2009/2/article/486*, accessed April 30, 2012.

Turetsky, M. R., E. S. Kane, J. W. Harden, R. D. Ottmar, K. Manies, E. Hoy, and E. Kasischke. 2010. Recent acceleration of biomass burning and carbon losses in Alaskan forests and peatlands. Nature Geoscience 4:27-31.

Turner, J., R. A. Bindschadler, P. Convey, G. D. Prisco, E. Fahrbach, J. Gutt, D. A. Hodgson, P. A. Mayewski, and C. Summerhayes. 2009. Antarctic Climate Change and the Environment: A Contribution to the International Polar Year 2007-2008. Cambridge, UK: Scientific Committee on Antarctic Research.

U.S. Department of the Interior. 2008. Endangered and Threatened Wildlife and Plants; Determination of Threatened Status for the Polar Bear (*Ursus maritimus*) Throughout Its Range. Final Rule. Federal Register 73(95):76249-76269.

U.S. Navy. 2009. U.S. Navy Arctic Roadmap. Available at *http://www.navy.mil/navydata/documents/USN_artic_roadmap.pdf*, accessed April 30, 2012.

Urban, K. D., A. J. Gerrard, Y. Bhattacharya, A. J. Ridley, L. J. Lanzerotti, and A. T. Weatherwax. 2011. Quiet time observations of the open-closed boundary prior to the CIR-induced storm of 9 August 2008. Space Weather 9:13.

Valencik, K. 2010. Sea level rise: Delaware's rising tide. Outdoor Delaware Fall 2010.

Vaughan, D. G., G. J. Marshall, W. M. Connolly, C. Parkinson, R. Mulvaney, D. A. Hodgson, J. C. King, C. J. Pudsey, and J. T. Vaughan. 2003. Recent rapid regional climate warming on the Antarctic Peninsula. Climate Change 60:243-274.

Velicogna, I. 2009. Increasing rates of ice mass loss from the Greenland and Antarctic ice sheets revealed by GRACE. Geophysical Research Letters 36:L19503.

Vermeer, M., and S. Rahmstorf. 2009. Global sea level linked to global temperature. Proceedings of the National Academy of Sciences of the United States 106(51):21527-21532.

Weiler, C. S., S. Drobot, and J. Baeseman. 2008. New Generation of Polar Researchers. Symposium (NGPR) La Foret Conference Center, Colorado Springs, CO, May 4-11, 2008.

Weller, F., M. Nolan, G. Wendler, C. Benson, K. Echelmeyer, and N. Untersteiner. 2007. Fifty years of McCall Glacier research: From the International Geophysical Year 1957-58 to the International Polar Year 2007-08. Arctic 60(1):101-110.

White, J. W. C., R. Alley, J. Brigham-Grette, J. Fitzpatrick, A. Jennings, S. Johnsen, G. Miller, R. S. Nerem, and L. Polyak. 2010. Past rates of climate change in the Arctic. Quaternary Science Reviews 29(15-16):1716-1727.

Wiebe, P. H., C. Ashjian, G. Lawson, and N. J. Copley. 2011. Horizontal and vertical distribution of euphausiid species on the Western Antarctic Peninsula U.S. GLOBEC Southern Ocean study site. Deep Sea Research Part II: Topical Studies in Oceanography 58(13-16):1630-1651.

Wilson, D. S., and B. P. Luyendyk. 2009. West Antarctic paleotopography estimated at the Eocene-Oligocene climate transition. Geophysical Research Letters 36:L16302.

Wilson, J. T. 1961. I.G.Y.: The Year of the New Moons. New York: Knopf.

Winther, G. 2010. The Political Economy of Northern Regional Development. Copenhagen: Nordic Council of Ministers.

Woodgate, R. A., T. Weingartner, and R. Lindsay. 2010. The 2007 Bering Strait oceanic heat flux and anomalous Arctic sea-ice retreat. Geophysical Research Letters 37:L01602.

Wu, X. P., M. B. Heflin, H. Schotman, B. L. A. Vermeersen, D. A. Dong, R. S. Gross, E. R. Ivins, A. Moore, and S. E. Owen. 2010. Simultaneous estimation of global present-day water transport and glacial isostatic adjustment. Nature Geoscience 3(9):642-646.

Xu, J. F., and D. J. Goering. 2008. Experimental validation of passive permafrost cooling systems. Cold Regions Science and Technology 53(3):283-297.

Young, T., and P. Bjerregaard, eds. 2008. Health Transitions in Arctic Populations. Toronto: University of Toronto Press.

Zhang, J., M. Steele, R. Lindsay, A. Schweiger, and J. Morison. 2008. Ensemble 1-year predictions of Arctic sea ice for the spring and summer of 2008. Geophysical Research Letters 35:L08502.

Zicus, S. 2011. IPY Education activities. In Understanding Earth's Polar Challenges: International Polar Year 2007-2008, I. KrupnikI. Allison, R. Bell, P. Cutler, D. Hik, J. López-Martínez, V. Rachold, E. Sarukhanian, and C. Summerhayes, eds. Edmonton, Canada: CCI Press.

# Appendixes

# Appendix A

## Statement of Task

International Polar Year (IPY) 2007-2008 was an intense, international campaign of polar observations, research, and analysis designed to further understanding of the polar regions. With the completion of the main, fieldwork phase of IPY, an ad hoc committee will produce a report that:

- highlights the outcomes (new scientific discoveries, observations, and findings, including infrastructure and education and outreach contributions) of the multifaceted IPY campaign from a U.S. perspective,
- integrates the lessons from different activities, including lessons learned about the benefits gained and challenges posed by international and multidisciplinary collaborations and by data access and management issues, and
- records U.S. IPY efforts so they are available to a broad audience including researchers, decision makers, and stakeholders.

This study will be based heavily on information generated at a large community workshop. It will look across disciplines and at both poles. The workshop will serve as a forum to facilitate community participation in a comprehensive synthesis of U.S. IPY efforts and accomplishments. The workshop will be organized to address four themes (introduced below and developed in detail by the appointed planning committee), including one specifically highlighting education and outreach activities. The workshop will feature invited presentations, discussions, and breakout group synthesis.

The committee will identify the major cross-cutting lessons of this IPY and discuss why these lessons are important today in planning for the future. It will explore "next steps"—how to keep the momentum gained in polar science during IPY to continue, especially given the increased relevance of the polar regions, and their importance to our understanding of climate change and adaptation strategies. It will consider whether the concept of holding large "international years" still holds value in times when international coordination is no longer a rare way to do research. In total, the workshop and study report will illustrate how the many pieces of IPY combine to move polar understanding forward in significant and sometimes unexpected ways.

# Appendix B

## Workshop on the Lessons and Legacies of International Polar Year 2007-2008: Agenda and Participant List

**June 15-16, 2011**
**Convened by the National Research Council**
**Polar Research Board**

**The National Conference Center (NCC)**
**18980 Upper Belmont Place Drive**
**Leesburg, VA 20176**

### AGENDA

**Wednesday, June 15, 2011—Loudon Room**

***7:15 am—Registration and Working Breakfast available Loudon Room***

| | |
|---|---|
| 8:00 | Welcome from NRC Committee Co-chairs Julie Brigham-Grette and Bob Bindschadler |
| 8:15 | Opening remarks—Karl Erb, National Science Foundation |

**Plenary Session: Discoveries** **(Moderator: Mary Albert)**

| | |
|---|---|
| 8:30 | *Stories from the Ice, Ice Sheet and Sea Ice Discoveries from IPY*<br>Don Perovich, ERDC-CRREL and Ted Scambos, NSIDC |
| 8:45 | *Oceanographic Activities During IPY 2007-2008: Key Discoveries and Future Activities*<br>Jackie Grebmeier, University of Maryland |
| 9:00 | *IPY Discoveries Within the Linked Spheres of Ice, Water, and Air: From Sea Ice to Stratosphere*<br>Cecilia Bitz, University of Washington and Terry Deshler, University of Wyoming |
| 9:15 | *History of Polar Climates: The IPY Contributions*<br>Richard Alley, Pennsylvania State University |
| 9:30 | Q&A |
| 10:00 | Break |

**Working Groups: Discoveries**

| | |
|---|---|
| 10:15 | 8 working groups each with assigned Moderator and Rapporteur<br>(Note: selected topics and questions to be addressed are forthcoming) |

***12:00-1:15 pm Working Lunch available***

**Plenary Session: Working Groups Report on Discoveries** **(Moderator: John Cassano)**

| | |
|---|---|
| 1:15 | 8 Moderator/Rapporteurs report back for 5 minutes each using a maximum of 2 slides |
| 1:55 | Open Q&A and discussion |

**Plenary Session: Tools of the IPY** **(Moderator: Larry Hinzman)**

2:30 *Satellites and Aircraft and the Transformation of Polar Paradigms*
Waleed Abdalati, National Aeronautics and Space Administration

2:45 *Observing Systems New and Old Facilitating Polar Research*
Peter Schlosser, Lamont-Doherty Earth Observatory, Columbia University

3:00 *Application of Genomics Tools Across Polar Biology*
Jody Deming, University of Washington

3:15 Q&A

***3:45 Break***

**Plenary Session: People** **(Moderator: Tom Taylor)**

4:00 *The People of POLAR-PALOOZA (aka "The Real Researchers of IPY")*
Geoffrey Haines Styles, POLAR-PALOOZA/P2K

4:15 *IPY EOC: Inspiring Projects Yielding Enthusiasm and Outstanding Capacity*
Jenny Baeseman, Association of Polar Early Career Scientists

4:30 Q&A

5:00 Visit draft report outline — Bob Bindschadler

5:30 Goals for evening and day 2 — Julie Brigham-Grette

**5:45 ADJOURN**

6:30 Reception followed by Working Dinner for all participants - West Terrace

*****Shuttle departs at 8:30 pm from the NCC for the Springhill Suites Hotel*****

**Thursday, June 16, 2011—Loudon Room**

***** Shuttle departs at 7:00 am and 7:15 am from the Springhill Suites Dulles to NCC*****

***7:15 am—Working Breakfast available Loudon Room***

8:00 Highlights from Day 1, Plan for the Day 2 — Julie Brigham-Grette and Bob Bindschadler

**Working Groups: Tools of the IPY and People**

8:15 4 working groups each with assigned Moderator and Rapporteur—Tools of the IPY
(Note: selected topics and questions to be addressed are forthcoming)
4 working groups each with assigned Moderator and RapporteurPeople of the IPY
(Note: selected topics and questions to be addressed are forthcoming)

**Plenary Session: Working Groups Report on Tools and People** **(Moderator: Eileen Hofmann)**

9:45 8 Moderator/Rapporteurs report back for 5 minutes each using maximum of 2 slides

10:25 Open Q&A and discussion

**Plenary Session: Knowledge to Action** **(Moderator: Stephanie Pfirman)**

10:45 *The Arctic: No Longer an Optional Ocean*
Mead Treadwell, Lt. Governor of Alaska

11:00 *From IPY Testbeds to Innovative Communities of Practice*
Hajo Eicken, University of Alaska Fairbanks

11:15 Open Q&A

***11:45-1:00pm Working Lunch available***

**Plenary Session: Reflections** **(Moderator: Igor Krupnik)**

1:00 *Leaders, Lessons and Legacies: The Sponsors' Global Perspective*
Carthage Smith, International Council for Science

1:15 *Icebreakers and Kayaks: Ideas of Many Sizes in This IPY*
Dave Carlson, UNAVCO

1:30 *Learning Other Languages*
Fae Korsmo, National Science Foundation

1:45 Open Q&A

**Working Groups: Knowledge to Action and Reflections**

2:00 4 working groups each with assigned Moderator and Rapporteur—Knowledge to Action
(Note: selected topics and questions to be addressed are forthcoming)
4 working groups each with assigned Moderator and Rapporteur—Reflections
(Note: selected topics and questions to be addressed are forthcoming)

***3:30*** ***Break***

**Plenary Session: Working Groups Report on K2A and Reflections** **(Moderator: Chris Rapley)**

3:45 8 Moderator/Rapporteurs report back for 5 minutes each using maximum of 2 slides

4:25 Open Q&A and discussion—Revisit Outline

5:00 **Closing Remarks** Bob Bindschadler and Julie Brigham-Grette
**ADJOURN**

**Friday, June 17, 2011Loudon Room**

***8:00 amWorking Breakfast available***

8:30 NRC IPY Committee meeting - Closed Session

**3:00** **ADJOURN**

## PARTICIPANTS

**Waleed Abdalati**, National Aeronautics and Space Administration
**Mary Albert**, Dartmouth College
**Richard Alley**, Pennsylvania State University
**Jenny Baeseman**, Association of Polar Early Career Scientists
**Bob Bindschadler**, National Aeronautics and Space Administration
**Sara Bowden**, International Arctic Science Committee
**Julie Brigham-Grette**, University of Massachusetts, Amherst
**John Cassano**, University of Colorado, Boulder
**Rob DeConto**, University of Massachusetts, Amherst
**Jody Deming**, University of Washington
**Terry Deshler**, University of Wyoming
**Hugh Ducklow**, The Ecosystems Center at Marine Biological Laboratory
**Hajo Eicken**, University of Alaska, Fairbanks
**Karen Frey**, Clark University
**Shari Gearhard**, University of Colorado, Boulder, National Snow and Ice Data Center
**Victoria Gofman**, Aleut International Association/Bering Sea Sub-Network (BSSN)
**Prasad Gogineni**, University of Kansas

**Jackie Grebmeier**, University of Maryland Center for Environmental Studies
**Geoffrey Haines-Stiles**, POLAR-PALOOZA/P2K
**William Hammer**, Augustana College
**David Hik**, University of Alberta, President IASC
**Larry Hinzman**, University of Alaska, Fairbanks, International Arctic Research Center
**Eileen Hofmann**, Old Dominion University
**Bernice Joseph**, University of Alaska, Fairbanks, College of Rural and Community Development
**Chuck Kennicutt**, Texas A&M University, President SCAR
**Fae Korsmo**, National Science Foundation
**Igor Krupnik**, Smithsonian Institution
**Roger Launius**, Smithsonian Institution
**Amy Lauren Lovecraft**, University of Alaska, Fairbanks
**Doug Martinson**, Columbia University Lamont-Doherty Earth Observatory
**Jeremy Mathis**, University of Alaska, Fairbanks
**Joe McConnell**, Desert Research Institute
**Walt Meier**, University of Colorado, Boulder, National Snow and Ice Data Center
**Flavio Mendez**, National Science Teachers Association
**Vera Metcalf**, Eskimo Walrus Commission
**Alison Murray**, Desert Research Institute
**Mark Parsons**, University of Colorado, Boulder, National Snow and Ice Data Center
**Donald Perovich**, U.S. Army ERDC-Cold Regions Research and Engineering Lab
**Stephanie Pfirman**, Barnard College
**Ross Powell**, Northern Illinois University
**Frank Rack**, University of Nebraska, Lincoln
**Chris Rapley**, University College London
**Vladimir Romanovsky**, University of Alaska, Fairbanks
**Hal Salzman**, Rutgers University
**Ted Scambos**, University of Colorado, Boulder, National Snow and Ice Data Center
**Peter Schlosser**, Columbia University, Lamont-Doherty Earth Observatory
**Mark Serreze**, University of Colorado, Boulder ,National Snow and Ice Data Center
**Carthage Smith**, International Council for Science
**Elena Bautista Sparrow**, University of Alaska, Fairbanks
**Lisa Speer**, Natural Resources Defense Council
**Sharon Stammerjohn**, University of California, Santa Cruz
**Michelle Toohey**, Deputy Chief of Staff to Mead Treadwell
**John Toole**, Woods Hole Oceanographic Institution
**Mead Treadwell**, Lieutenant Governor of Alaska
**Craig Tweedie**, University of Texas, El Paso
**Isabella Velicogna**, University of California, Irvine
**Ross Virginia**, Darmouth College
**Diana Wall**, Colorado State University
**Katie Walter**, University of Alaska, Fairbanks
**Allan Weatherwax**, Siena College
**Wilford Weeks**, University of Alaska, Fairbanks
**James White**, University of Colorado, Boulder, Institute of Arctic and Alpine Research
**Doug Wiens**, Washington University, St. Louis
**Terry Wilson**, The Ohio State University

# Appendix C

## Biographical Sketches of Committee Members

**Julie Brigham-Grette** (*Co-chair*) is a professor in the Department of Geosciences at the University of Massachusetts, Amherst. Dr. Brigham-Grette received her Ph.D. from the University of Colorado's Institute for Arctic and Alpine Research. After postdoctoral research at the University of Bergen, Norway, and the University of Alberta, Canada, with the Canadian Geological Survey, she joined the faculty at the University of Massachusetts in the fall of 1987. Dr. Brigham-Grette has been conducting research in the Arctic for nearly 34 years, including nine field seasons in remote parts of northeast Russia since 1991. Her research interests and experience span a broad spectrum dealing with Arctic paleoclimate records and the Late Cenozoic evolution of the Arctic climate both on land and offshore, especially in the Bering Strait region. She was a member of the Arctic Logistics Task Force for the National Science Foundation (NSF) Office of Polar Programs (OPP) 1996-1999 and 2000-2003, and was member of the OPP Office Advisory Council 2002-2004. She chaired the U.S. Scientific Delegation to Svalbard for Shared Norwegian/U.S. Scientific Collaborations and Logistical Platforms in 1999. Brigham-Grette was two-term chair of the International Geosphere/Biosphere Program's Science Steering Committee on Past Global Change (PAGES) with an international program office in Bern, Switzerland, and past president of the American Quaternary Association. She served as one of two U.S. representatives to the International Continental Drilling Program. She is currently chair of the American Geophysical Union's Paleoclimate and Paleoceanography Focus Group and co-chair of the DOSECC Science Planning Committee for scientific drilling.

**Robert A. Bindschadler** (*Co-chair*)**, NASA Goddard Space Flight Center (Emeritus)** has been an active Antarctic field researcher for the past 30 years. He has led 15 field expeditions to Antarctica and has participated in many other expeditions to glaciers and ice caps around the world. He maintains an active interest in the dynamics of glaciers and ice sheets, primarily on Earth, investigating how remote sensing can be used to improve our understanding of the role of ice in the Earth's climate and exploring the forces driving ice sheet change. Applications developed by Dr. Bindschadler include measuring ice velocity and elevation using both visible and radar imagery, monitoring melt of and snowfall on ice sheets by microwave emissions, and detecting changes in ice sheet volume by repeat spaceborne altimetry. He has advised the U.S. Congress and the Vice President on the stability of ice sheets and ice shelves, led the West Antarctic Ice Sheet Initiative for 20 years, served on many scientific commissions and study groups as an expert in glaciology and remote sensing of ice, was instrumental in the planning of the International Polar Year, and is a past president of the International Glaciological Society. Some of the more significant awards he has received are: Goddard Award of Merit (2008), Fellow of the American Geophysical Union (2001), Goddard Senior Fellow (2000), Excellence in Federal Career (1989), the Antarctic Service Medal (1984), and the NASA Exceptional Scientific Achievement Medal (1994). He has published over 140 scientific papers and numerous review articles and has appeared on television and been heard on radio commenting on glaciological impacts of the climate on the world's ice sheets and glaciers.

**Mary R. Albert, Dartmouth College,** is professor of engineering at the Thayer School of Engineering at Dartmouth College, and she is executive director of the U.S. Ice Drilling Program Office. She was formerly a senior research scientist at the Army's Cold Regions Research and Engineering Lab. Her research includes heat, mass, chemical transfer, and electromagnetic processes in snow and firn, including atmosphere-snow exchange, ice core interpretation, and remote sensing of snow and ice. She has led and participated in many research programs in both Greenland and Antarctica, most recently as chief scientist of the Norwegian-U.S. Scientific Traverse of East Antarctica, an IPY project. While serving on the National Academies of Science Polar Research Board from 2003-2006, she was chair of the U.S. National Committee for the IPY and led the writing of the 2004 NRC Report, *A Vision for the International Polar Year*. Dr. Albert served on the NSF OPP Advisory Committee from 1998 to 2001, and was Chair of that committee from 1999 to 2000. She is currently associate editor of *Water Resources Research* and serves on the Executive Committee of the American Geophysical Union Cryosphere Focus Group. Dr. Albert earned her Ph.D. in Applied Mechanics and Engineering Sciences in 1991 from the University of California, San Diego.

**John Cassano, University of Colorado,** is an associate professor in the Department of Atmospheric and Oceanic Sciences and a Fellow of the Cooperative Institute for Research in the Environmental Sciences at the University of Colorado, Boulder. His research focuses on the meteorology and climate of the polar regions. Dr. Cassano is a U.S. delegate to the International Arctic Sciences Committee. Dr. Cassano received his Ph.D in Atmospheric Science from the University of Wyoming in 1998.

**Larry D. Hinzman, University of Alaska, Fairbanks,** is the director of the International Arctic Research Center and is professor of civil and environmental engineering at the University of Alaska, Fairbanks. Professor Hinzman's primary research interests involve permafrost hydrology. He has conducted hydrological and meteorological field studies in the Alaskan Arctic continuously for over 30 years while frequently collaborating on complementary research in the Russian and Canadian Arctic. His research efforts have involved characterizing and quantifying hydrological processes and their interdependence with climate and ecosystem dynamics. Dr. Hinzman's academic degrees were earned from South Dakota State University, Purdue University, and the University of Alaska, Fairbanks in chemistry, soil science, agronomy and soil physics. He has served as a member of the U.S. Polar Research Board, the U.S. Representative to the International Permafrost Association and is a member of the Universities Council on Water Resources. He served as co-chair of the U.S. National Science Foundation study on the Arctic Freshwater Initiative and presently serves as chief scientist for the U.S. Department of Energy Arctic Next Generation Ecosystem Experiment. He is an internal advisory committee member for the Alaska Center for Energy and Power and Association of Polar Early Career Scientists. Dr. Hinzman serves on the International Advisory Board of the Korean Polar Research Institute and is strongly committed to facilitating international partnerships to advance our understanding of the Arctic system.

**Dr. Eileen E. Hofmann, Old Dominion University,** is a professor of iceanography in the Department of Ocean, Earth and Atmospheric Sciences and a member of the Center for Coastal Physical Oceanography, both at Old Dominion University. Dr. Hofmann earned a Ph.D. in Marine Science and Engineering from North Carolina State University. Her research interests are in the areas of understanding physical-biological interactions in marine ecosystems, climate control of diseases of marine shellfish populations, descriptive physical oceanography, and mathematical modeling of marine ecosystems. She has worked in a variety of marine environments, most recently the continental shelf region off the western Antarctic Peninsula. She served on the Ocean Studies Board and on numerous National Research Council committees, including the Committee on Strategic Advice on the U.S. Climate Change Science Program. She is currently the chair of the Integrated Marine Biogeochemical and Ecosystem Research Project, cosponsored by the

International Geosphere-Biosphere Program and the Scientific Committee for Oceanic Research.

**Igor Krupnik, Smithsonian Institute,** is curator of Arctic and Northern Ethnology collections at the National Museum of Natural History, Smithsonian Institution, in Washington, D.C. His primary research fields are modern cultures, ecological knowledge, and cultural heritage of the people of the Arctic, primarily in Alaska and Siberia; culture change and contact history; human ecology; history of Arctic science and Arctic indigenous studies; and impact of modern climate change on Arctic residents, their economies, and cultures. Dr. Krupnik served on the U.S. National Planning Committee for IPY in 2003-2004, before being nominated to the main international steering body for IPY, the ICSU-WMO Joint Committee, in 2004. On the Joint Committee (2005-2010), Dr. Krupnik served as one of two social scientists representing the interests of social studies and Arctic residents. He was instrumental in bringing social/human research onto the IPY agenda. Dr. Krupnik's personal contribution to IPY science program was an international project called SIKU (Sea Ice Knowledge and Use in the North), on which he coordinated activities of several research teams from Canada, the United States, Russia, Greenland, and France that worked in some 30 Arctic communities from the Bering Strait to Greenland. He was the lead editor of the main summary report on IPY activities, "Understanding Earth's Polar Challenges: International Polar Year 2007-2008," by the IPY Joint Committee (2011). Dr. Krupnik received his Ph.D. in anthropology from the Institute of Ethnology, Russian Academy of Sciences.

**Vera Kingeekuk Metcalf, Marine Mammal Commission,** is the director of the Eskimo Walrus Commission (EWC) at Kawerak, Inc. since 2002. She continues to work in promoting local community participation in research that involves a community's natural and cultural resources. In 2004 and in cooperation with U.S. Fish and Wildlife Service, EWC convened a workshop to discuss and begin integrating research concerns with the Pacific walrus and its environment. As EWC director, Ms. Metcalf also serves as a Special Advisor on Native Affairs–Marine Mammal Commission, the Pacific Walrus Technical Committee, and on the Pacific Walrus Conservation Fund. Ms. Metcalf also represents EWC as an Advisory Panel member on the North Pacific Research Board and on the Indigenous People's Council on Marine Mammals (consisting of commissions formed to identify and address marine mammal issues of common concerns). She is currently serving on the Inuit Circumpolar Council Alaska and its Executive Committee. Ms. Metcalf is a former commissioner for the U.S. Arctic Research Commission.

**Stephanie Pfirman, Barnard College,** is Alena Wels Hirschorn '58 and Martin Hirschorn Professor in Environmental and Applied Sciences and co-chair of the Department of Environmental Science at Barnard College, which she joined in 1993. She holds a joint appointment with Columbia University where she is a member of the faculties of the Earth Institute and the Department of Earth and Environmental Sciences, and adjunct research scientist at the Lamont-Doherty Earth Observatory of Columbia University. Throughout her career, Pfirman has been involved with researching the Arctic environment, undergraduate education, environmental policy strategies, and public outreach. Current interests include environmental aspects of sea ice in the Arctic, climate change education, and the development of women scientists and interdisciplinary scholars. In 2010, Pfirman was elected a fellow of the American Association for the Advancement of Science "for distinguished contributions to scientific studies of the Arctic and effective outreach to policy makers, students, faculty and the general public." The first chair of NSF's Advisory Committee for Environmental Research and Education (ACERE), Dr. Pfirman oversaw analysis of a 10-year outlook for environmental research and education. Dr. Pfirman rejoined the ACERE in 2010, and she also is currently a member of NSF's Merit Review Process Advisory Committee. She is a past member of the National Academy of Sciences Polar Research Board, which served as the U.S. National Committee for the International Polar Year 2007-2009, past president of the Council of Environmental Deans and Directors, and past chair of NSF's Office Advisory Committee to the Office of Polar Programs. Dr. Pfirman earned her Ph.D. from the Massachusetts Institute of Technology/Woods Hole Oceanographic Institution Joint Program in Oceanography and Oceanographic Engineering.

**Chris Rapley, The Science Museum,** is professor of Climate Science at University College London (UCL). He earned an M.Sc. in Radio Astronomy at Jodrell Bank in Cheshire followed by a Ph.D. at the Mullard Space Science Laboratory (MSSL) at University College London on the origin of the cosmic soft X-ray diffuse background. Following a decade as the founder and head of the Earth Observation group and associate director at UCL's Mullard Space Science laboratory. Professor Rapely was appointed Executive Director of the International Geosphere-Biosphere Programme IGBP, which he ran from 1994 to 1998. He was director of the British Antarctic Survey from 1998 to 2007 during which time he was a vice president then president of the Scientific Committee on Antarctic Research (SCAR) and the chair of the planning group that developed the International Polar Year 2007-2008. He was director of the Science Museum from 2007 to 2010, during which time the Museum delivered its Centenary programme, including the new gallery "Atmosphere: Exploring Climate Science." In 2008 he was awarded the Edinburgh Science Medal for "professional achievements judged to have made a significant contribution to the understanding and well-being of humanity."

**Lisa Speer, Natural Resources Defense Council,** is the director of the International Oceans Program at NRDC, an environmental organization dedicated to protecting natural resources and public health with offices in the United States and China. Her work currently focuses on conservation and management of the Arctic marine environment, and marine biodiversity beyond national jurisdiction, an area known as the "high seas." Ms. Speer conducts advocacy in a variety of international forums to promote integrated, ecosystem-based management of human activities on the high seas and in the Arctic, with a particular focus on marine fisheries. She received her Master's degree from Yale University and her Bachelor's degree from Mount Holyoke College. Ms. Speer has served as a member of the NRC Board on Environmental Studies and Toxicology, and on ad hoc NRC study committees.

**Thomas N. Taylor, University of Kansas,** is Roy A. Roberts Distinguished Professor at the University of Kansas. He is also senior curator of the Natural History Museum and Biodiversity Research Center, and courtesy professor for the Department of Geology. He also serves as director of the State of Kansas NSF EPSCoR Program. He earned his Ph.D. in botany and geology from the University of Illinois in 1964. Dr. Taylor is a member of the National Academy of Sciences. He also serves on the National Science Foundation Education and Human Resources Advisory Committee, as chair of the Strategic Planning and Assessment Committee for National Institutes of Health BRIN KU Medical Center, on Senator Pat Roberts' Advisory Committee in Science, Technology, and on the Future Kansas Implementation Advisory Committee, the National Science Foundation GPRA Performance Assessment Advisory Committee, the National Science Foundation MPSAC/EHRAC Committee to Review Undergraduate Education in Math and the Physical Sciences, Bioinformatics Core Advisory Committee. He serves on multiple NSF EPSCoR Advisory Boards and committees. He served on the Polar Research Board for the NRC. In addition he served as faculty advisor to the chancellor of the Ohio Board of Regents and on the Government-University-Industry Research Roundtable for the State of Ohio.

**Wilford F. Weeks, University of Alaska, Fairbanks,** is professor emeritus of geophysics at the University of Alaska. His primary area of interest is in the properties and geophysical behavior of the sea ice covers of the world's oceans. Specific areas he has investigated include interrelations between growth conditions and the structure, composition, and mechanical and electromagnetic properties of sea ice; formation and statistical characteristics of pressure ridges; ice-induced gouging of the sea floor, bearing capacity and forces exerted by moving ice; and application of varied remote sensing techniques to sea ice problems and general problems relating to atmosphere-ice-ocean interactions. Dr. Weeks is a member of the National Academy of Engineering. He has also had considerable experience concerning the geophysics and engineering of snow and ice masses in general, including the structure of lake and river ice, winter heat loss from rivers, avalanche forecasting, properties of alpine snow, and temperature distributions and snow property variations in central Greenland. Dr. Weeks received his Ph.D. from the University of Chicago.

# Appendix D

## Acronyms and Initialisms

| | |
|---|---|
| ACCE | Antarctic Climate Change and the Environment |
| AFHCAN | Alaska Federal Health Care Access Network |
| AG | Action Group (SCAR) |
| AGAP | Antarctic Gamburtsev Province/ Antarctica's Gamburtsev Province Project |
| AGU | American Geophysical Union |
| AHHI | Arctic Human Health Initiative |
| ANDRILL | ANtarctic Geological DRILLing |
| AOGCM | Atmosphere-Ocean General Circulation Model |
| APECS | Association of Polar Early Career Scientists |
| ARMAP | Arctic Research Mapping Application |
| ASAID | Antarctic Snow Accumulation and Ice Discharge |
| ASEP | Amundsen Sea Embayment Project |
| ASI | Arctic Social Indicators |
| ASR | Arctic System Reanalysis |
| ASTIS | Arctic Science and Technology Information System |
| ATCM | Antarctic Treaty Consultative Meeting |
| BSSN | Bering Sea Sub-Network |
| CADIS | Cooperative Arctic Data & Information Service |
| CAML | Census of Antarctic Marine Life |
| CANHR | Center for Alaska Native Health Research |
| CASIE | Characterization of Arctic Sea Ice Experiment |
| CAVIAR | Community Adaptation and Vulnerability in the Arctic Region |
| CBMP | Circumpolar Biodiversity Monitoring Program |
| CCAMLR | Conservation of Antarctic Marine Living Resources |
| CDOM | colored dissolved organic matter |
| CIRES | Cooperative Institute for Research in Environmental Sciences |
| CliC | Climate and Cyrosphere |
| CODATA | Committee on Data for Science and Technology |
| COMPASS | Comprehensive Meteorological Data Set of Active IPY Antarctic Measurement Phase for Scientific and Applied Studies |
| CRBP | Cold Regions Bibliography Project |
| DADDI | Discovery, Access, and Delivery of Data for IPY |
| DAHLI | Discovery and Access of Historic Literature of the IPYs |
| DAMOCLES | Developing Arctic Modeling and Observing Capabilities for Long-Term Environmental Studies |
| DISC | Deep Ice Sheet Coring |
| DOC | dissolved organic carbon |
| DOI | U.S. Department of the Interior |

| | |
|---|---|
| EAIS | East Antarctic ice sheet |
| EGU | European Geosciences Union |
| eGY | Electronic Geophysical Year |
| ELOKA | Exchange of Local Knowledge in the Arctic |
| EPA | U.S. Environmental Protection Agency |
| EPB | European Polar Board |
| EPPR | emergency preparedness and response mechanisms |
| ESA | Endangered Species Act |
| FWS | U.S. Fish and Wildlife Service |
| GCM | General Circulation Model |
| GCMD | Global Change Master Directory |
| GIIPSY | Global Inter-agency IPY Polar Snapshot Year |
| GLOBE | Global Learning and Observations to Benefit the Environment |
| GRACE | Gravity Recovery and Climate Experiment |
| HHS | U.S. Department of Health and Human Services |
| IACP | Arctic Institute for Applied Circumpolar Policy |
| IAGA | International Association of Geomagnetism and Aeronomy |
| iAOOS | integrated Arctic Ocean Observing System |
| IASC | International Arctic Science Council |
| IASOA | International Arctic Systems for Observing the Atmosphere |
| ICC | Inuit Circumpolar Council |
| ICDP | International Continental Drilling Program |
| ICS | International Circumpolar Surveillance |
| ICSU | International Council for Science |
| IDDO | Ice Drilling Design and Operations |
| IGERT | Interdisciplinary Graduate Education, Research, and Training |
| IGS | International Glaciological Society |
| IGY | International Geophysical Year |
| IPCC | Intergovernmental Panel on Climate Change |
| IPO | International Programme Office |
| IPY | International Polar Year (IPY 2007-2008?) |
| IPYDIS | IPY Data and Information System |
| IPYPD | International Polar Year Publications Database |
| ISAC | International Study of Arctic Change |
| IUCH | International Union for Circumpolar Health |
| IUGG | International Union of Geodesy and Geophysics |
| JC | Joint Committee |
| LASP | Laboratory for Atmospheric and Space Physics, University of Colorado |
| LIMA | Landsat Image Mosaic of Antarctica |
| MaxNDVI | Maximum Normalized Difference Vegetation Index |
| MEOP | Marine Mammals as Explorers of the Ocean Pole to Pole |
| MIS | McMurdo Ice Shelf |
| MMCO | Mid-Miocene Climatic Optimum |
| NAS | National Academy of Sciences |
| NASA | National Aeronautics and Space Administration |
| NCAR | National Center for Atmospheric Research |
| NEEM | North Greenland Eemian Ice Drilling |
| NF | Northern Forum |
| NGPR | Next Generation Polar Research |
| NGRIP | North Greenland Ice Core Project |
| NISC | National Information Services Corporation |
| NOAA | National Oceanic and Atmospheric Administration |
| NPS | National Park Service |
| NRC | National Research Council |
| NSF | National Science Foundation |
| NSIDC | National Snow and Ice Data Center |

| | |
|---|---|
| NSTA | National Science Teachers Association |
| OAIPMH | Open Archives Initiative Protocol for Metadata Harvesting |
| OCB | open-closed boundary (magnetosphere) |
| ONR | Office of Naval Research |
| OPP | National Science Foundation Office of Polar Programs |
| POLARCAT | POLar study using Aircraft, Remote sensing, surface measurements and modeling of Climate, chemistry, Aerosols and Transport |
| PolarTREC | Teachers and Researchers Exploring and Collaborating |
| POLENET | Polar Earth Observing Network |
| POP | persistent organic pollutant |
| PRB | U.S. Polar Research Board |
| ROAM | Research and Educational Opportunities in Antarctica for Minorities |
| RVIB | Research Vessel/Icebreaker |
| SACNAS | Society for the Advancement of Chicanos and Native Americans in Science |
| SAON | Sustaining Arctic Observing Networks |
| sar | Synthetic Aperture Radar |
| SCAR | Scientific Committee on Antarctic Research |
| SCSCS | Spitsbergen Climate System Current Status |
| SEAOS | Southern Elephant Seals as Oceanographic Sensors |
| SEARCH | Study of Environmental Arctic Change |
| SHEBA | Surface Heat Budget of the Arctic Ocean Project |
| SIERRA | Sensor Integrated Environmental Remote Research Aircraft |
| SIKU | Sea Ice Knowledge and Use |
| SIMBA | Sea Ice Mass Balance in the Antarctic |
| SIPEX | Sea Ice Physics and Ecosystem eXperiment (Australia) |
| SIWO | Sea Ice for Walrus Outlook |
| SO GLOBEC | Southern Ocean Global Ocean Ecosystems Dynamics |
| SOOS | Southern Ocean Observing System |
| SPARC-IPY | Stratospheric Processes And their Role in Climate-International Polar Year |
| SPRI | Scott Polar Research Institute |
| UAS | unmanned aerial systems |
| UNCLOS | United Nations Convention on the Law of the Sea |
| UNFCCC | United Nations Framework Convention on Climate Change |
| USCG | U.S. Coast Guard |
| USGS | U.S. Geological Survey |
| WAIS | West Antarctic Ice Sheet |
| WAP | Western Antarctic Peninsula |
| WMO | World Meteorological Organization |